INTRODUCTION TO
HIGH POWER PULSE TECHNOLOGY

ADVANCED SERIES IN ELECTRICAL AND COMPUTER ENGINEERING

Editor: W. K. Chen

Advanced Series in Electrical and Computer Engineering – Vol. 10

INTRODUCTION TO
HIGH POWER PULSE TECHNOLOGY

S T Pai
Department of Electrical Engineering
Tsinghua University

Qi Zhang
Institute of Electronics
Academia Sinica

World Scientific
Singapore • New Jersey • London • Hong Kong

Published by

World Scientific Publishing Co. Pte. Ltd.
P O Box 128, Farrer Road, Singapore 9128
USA office: Suite 1B, 1060 Main Street, River Edge, NJ 07661
UK office: 57 Shelton Street, Covent Garden, London WC2H 9HE

INTRODUCTION TO HIGH POWER PULSE TECHNOLOGY

ISBN 981-02-1714-5

This book is printed on acid-free paper.

Printed in Singapore by Uto-Print

Preface

As demanding of power levels in various researches and developments, e.g. the developments of inertial confinement fusion, high power particle beams, intense lasers and so on, is getting higher and higher, high power pulse technology has recently become one of the most intensively studied subjects. This technology was originally initiated in the U.K. and the former USSR in the early 1960s. Since then a significant amount of progress has been made as a result of extensive research and development carried out by many technologically advanced countries, particularly by the USA and the former USSR. In spite of its long history of development and significant progress made in recent years, the literature concerning this technology remains rather diffuse. At present, there is no systematically written and well documented book which deals with the entire subject in existence. It is therefore very desirable to have a book to fill this blank, i.e. to have a book dealing with the entire technology and containing both the essential fundamentals as well as the various new developments. On this basis, the present authors intend to pull together the various literature and information concerning the subject into an introductory book. It is hoped that such an effort will provide some useful information to those who are interested in or working in this field.

This book is designed primarily to meet two objectives. It may serve as a text book of a semester university course for graduate or senior undergraduate students in the physical sciences, electrical engineering and other related disciplines or it may be used as a reference book for those who are working in the field. For those who wish to use the book for self-study purposes, some prerequisites are advised. These should include a general background of medium level electromagnetism, electrical circuitry and some knowledge on plasma and discharge physics. In order to meet these diverse objectives, the authors have made efforts to make the present book reasonably compact so that it could fit into a one semester schedule meanwhile retaining its comprehensiveness when serving as a reference book. The contents of the book are so arranged that theory and practice are proportionally balanced and each topic shall consist of essentially four basic elements: (1) fundamental principle, (2) mathematical expressions and formulas, (3) examples and illustrations, (4) numerical data and applications. In order to keep its compactness, lengthy theoretical discussions and detailed mathematical derivations are to be avoided whenever possible. For those who are interested in the details or wish to study the subject in depth, they are suggested to refer to the appropriate references given at the end of each chapter.

The authors have put every effort in making the book up-to-date and reasonably comprehensive. However, as this technology advances so fast in addition to the fact that the authors' limited ability in collecting these information at the time of writing, omission of some information may not be totally avoidable.

One of the authors, S. T. Pai would like to take this opportunity to express his appreciation for the financial support received from the National Natural Science Foundation of China.

CONTENTS

INTRODUCTION TO
HIGH POWER PULSE TECHNOLOGY

CHAPTER 1
INTRODUCTION

1-1 Background

On the basis of power levels involved, pulsed power technology may be loosely divided into two branches: low power pulse technology and high power pulse technology. The former mainly deals with the aspects concerning communication, high speed electronics, diagnostics and so on. There are several excellent books in these areas[1,2]. Interested readers are suggested to refer to these publications. In this book, we shall concentrate to discuss only those subjects which involve pulsed power in the megawatt (MW) or greater ranges. Technology dealing with these latter subjects are usually referred to as high power pulse technology (HPPT). In more specific terms, the typical ranges of the physical quantities frequently encountered in the HPPT are as follows.

Table 1-1 Typical Ranges of Quantities Involved

Energy	$(10^1 - 10^7)$ Joules
Power	$(10^6 - 10^{14})$ Watt
Voltage	$(10^3 - 10^7)$ Volt
Current	$(10^3 - 10^7)$ Amp
Current density	$(10^6 - 10^{11})$ Amp/M^2
Pulse width	$(10^{-10} - 10^{-5})$ Sec

As mentioned before, development of the high power pulse technology is largely the consequence of growing demands for high power source and for effective means of transferring the energy to the target. One typical example is the inertial confinement fusion (ICF) research project[3]. To achieve controlled thermonuclear fusion in the laboratory, a tiny target of D-T gas needs to be compressed to the density greater than $10^{23}/cm^3$ which requires particle or laser beams having power in the terawatt (TW) range and capable of transporting their energy to the target at a power deposition level greater then 10^{16} W/g[4]. Particle beams in the terawatt range cannot be produced with the conventional high energy particle accelerators as such accelerators can only supply current in the mA to A range, though the attainable particle energy can be as high as several hundred GeV. Deposition power achieved by means of conventional technologies is also far below the required level of 10^{16} W/g. Some new technology capable of meeting these requirements is needed. Development of the high power pulse technology is largely a consequence of such demands.

Besides inertial confinement fusion, a number of other developments have been also part of the driving force that further accelerated the developments of this new technology. The most notable one probably is the research for directed energy weapons, a military program conducted by the USA[5], perhaps a similar one by the former USSR[6]. Such program involves the research and development of chemical laser, excimer laser, x-ray laser, free-electron laser, neutral and charged particle beams, electromagnetic launchers, microwave and γ-ray generators, nuclear explosion simulators and so on[7-9]. A large part of these developments requires advanced technologies and special techniques, yet most of them do not exist at the time of investigation. As a result, a strong and immense technological force consisting of scientists and engineers mainly from the USA and other industrially advanced countries was thus formed, and with this the development of the high power pulse technology was pushed to a new frontier.

Other than the ICF research and the directed-energy weapons program, applications of the high power pulse technology in other areas are rather diverse. These may include magnetic confinement fusion, impact fusion, high energy physics research, plasma physics research, synchrotron radiation production, pulsed x-ray and γ-ray productions, projectile launching, metal parts forming, large area metal welding, radiographical diagnostics, gas clean up, materials processing and so on. All of these require some form of pulsed power. For instance, in the magnetic confinement fusion, hot plasma needs to be created, confined and heated. Each of these processes requires a huge amount of energy and short time of duration to complete it. That implies high power in pulsed form.

Development of a new technology usually requires well developed technologies already in existence as its foundation. There is no exception for the development of high power pulse technology. Its development is largely built up on the basis of the disciplines of high voltage electrical engineering and applied physics. From the fundamental point of view, it is hard to define a clear boundary which would separate the high power pulse technology from the last two disciplines. Alternatively one may well treat the present subject as a branch of high voltage electrical engineering or applied physics or a mixture of the two. Probably the last one is more close to the reality. For this reason, individuals who have sufficient training both in high voltage electrical engineering and applied physics may find it more advantageous in working in this field.

1-2 The Basic Characteristics of HPPT

Conventionally electric power is supplied in continuous forms of low power levels. Electric energy is delivered to the load slowly and steadily. For example, when we turn on a TV set, electric energy is continuously supplied from the power plant. In one evening's operation, say 3 hours, it may consume 1 million Joule of energy. However, in terms of power that is rather low, no more than a hundred watts or so, because power is defined as $P = \epsilon/t$ and in this case the time duration t is rather long (10800 sec) though the energy ϵ is considerably high. In many special applications, such as in the ICF research and directed-energy weapons program, energy must be delivered to the load very rapidly. Instead in continuous form, the power is required to be in the form of short pulses or bursts. The basic idea in the ICF process is to heat the D-T fuel very rapidly that an appreciable number of fusion reactions take place before it has a chance to blow itself apart. Based on estimates, heating a small D-T fuel pellet of 1 mm radius to fusion temperature, the required energy is about 1 million joules which is not really too great, roughly equivalent to the energy consumption in one evening's operation of a TV set. However when this energy is delivered to the fuel, say in 10^{-8} sec, it corresponds to a power of 100 TW. This represents a staggering power demand when comparing it with the maximum capacity of the world's largest hydro power plant, Itaipu of Brazil/Paraguay which is only about 0.01 TW. How is it possible to achieve such tremendous power? Certainly it cannot be done by conventional methods, whether electrical, chemical or mechanical. One has to rely on some new method i.e. the high power pulse technology.

High power pulse technology is a subject of unusual characters. In terms of experimentation, it is quite a complex and advanced technology. It involves very sophisticated facilities and requires highly specialized techniques to carry out the experiments. On the other hand, the basic principles to which it is based on are rather simple. The essential idea of the HPPT is that energy is collected from some primary energy source at low power level, low power density and stored in some temporary storage. Then the energy is rapidly released from storage and converted into a power of pulsed form. After further compression of the pulsed power, the power finally is delivered to the load at high power level and high power density. Fig.1-1 is a block diagram showing the basic structure and operating principle for a typical high power pulse facility. How is it possible to achieve such goals effectively and economically is the heart of this technology.

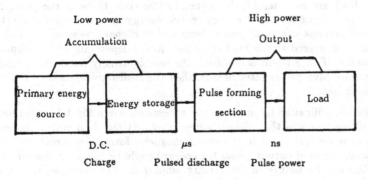

Fig.1-1 Block diagram of a typical high power pulse system

From Fig.1-1 we can see that, first of all, some kind of primary energy source must be available. This could be either electrical, chemical, mechanical or other forms of energy. However, in many applications, electrical energy, such as that supplied from an electrical generator, is found to be most appropriate. Energy from the primary source is usually slowly accumulated and stored. Energy storage can be done in many different ways. Energy can be stored in mechanical spring and flywheels, compressed gas, electric field (capacitors), magnetic field (inductors), electrochemical battery and so on. Among these techniques, electrical energy stored in capacitors and magnetic energy stored in inductors have been widely employed. Capacitors bank and Marx generator are two of the most widely used capacitive devices for such purpose. Both are made of a large number of high voltage capacitors and these capacitors are charged (accumulating energy) in parallel configurations. During discharge (releasing energy), the configuration of the former remains unchanged but the latter changes to series connection. Thus a Marx generator not only serves as an energy storage but also functions as a voltage multiplier, whereas a capacitor's bank can only serve as an energy storage. Transformation of the stored static energy into pulsed form is usually done by means of a switch between the energy store and the next section. Upon closing the switch the stored energy is transformed into a high power of pulsed form.

The pulse shape of the power is closely related to the performance of the switch. Requirements on the switch in such applications are rather stringent. Firstly, the switch must be able to withhold high voltages before closing. Secondly upon triggering, it must instantly react and transfer itself from a perfect insulator into a good conductor. Finally it must have the ability to recover from a conducting state to an insulating state quickly. The pulse forming section generally consists of a specially designed transmission line and some high voltage switches. Its main function is to further enhance the power level of the output power coming from the storage section by re-shaping the pulse form. Another function is to transmit the energy from the energy source to the load, as in some cases the energy source and the load are considerably far apart. After gone through the pulse forming section, the originally stored energy in the energy source is now converted into an electromagnetic energy of pulsed form and at higher power levels. This energy now can be delivered to the load of various forms depending on the applications. For example, if an *e*-beam is required, the load is usually an electron diode. If it is to produce high power laser beams, then the load represents the laser medium of the laser head.

Power amplification by means of the scheme shown in Fig.1-1, could be as high as 10 million times of the original power level. At the final stage, the energy or the power is generally still in the electromagnetic form. Upon further conversion in the load, the final output power (or energy) could be in one of the various forms depending on the nature of the facility employed. For example, in the particle beam fusion accelerators, the final output is in the form of ion or electron beam whereas in the intense laser cases, laser radiation will be the final output. In Fig.1-1, for simplicity we have omitted the switches there. In a real situation, there are switches between two blocks and these switches serve several important functions for stage isolation, energy transmission and pulse forming, etc. As mentioned before, the performance of the entire system is closely related to the characteristics and capability of the switches employed. In addition, not shown in Fig.1-1, there are also the diagnostics, controls, data acquisition, etc. These are all essential elements to a functional high power pulse system. In Fig.1-1 only the very basic structure and the fundamental procedure were shown. In the following chapters, we shall discuss each element of the system as well as the entire procedure in fine details, respectively.

1-3 Plan of Presentation

In Chapter 2, we shall devote the entire chapter to the topic of energy storage in which both the capacitive energy storage system and the inductive energy storage system shall be systematically discussed. In the discussion, the basic principle as well as the practical application shall be equally emphasized. Insulation is one of the crucial areas that need to be treated properly. Without proper insulation, no high voltage can be maintained. Full discussion of this subject shall be given in Chapter 3. It shall include insulations under gaseous, liquid, solid and vacuum environments as well as surface flashover at different surfaces and under various physical conditions. Insulation under pulsed voltages shall be particularly emphasized in the discussion, as such information is essential for pulse application but not generally available. Chapter 4 shall deal with the basic principle and application of pulse forming lines. Four types of transmission lines, namely the simple transmission line. Blumlein transmission line, multistage pulse forming line and the magnetically insulated transmission line shall be respectively discussed. Switching shall be discussed in Chapter 5. As pointed out before, switching is another crucial element in high power pulse system, therefore the various switching techniques shall all be discussed. These shall cover the respective areas for gaseous, liquid, solid, plasma and magnetic switches. Discussion of the open switching technique shall not be included in this chapter as such a topic is rather closely associated with the inductive energy storage system hence it is more appropriate to discuss it in Chapter 2. Chapter 6 shall deal with diodes and particle beams. Topics concerning the basic elements of electron diode, ion diode and particle transport techniques are to be discussed. After each of these basic elements mentioned above has been fully discussed, then we shall condense in Chapter 7 all the previous topic into a comprehensive one and use it for the discussion of the complete pulsed power system such as the particle beam accelerators, laser drivers, railgun and other high power pulse facilities. In the final chapter, i.e. Chapter 8, we shall devote to the discussion of various diagnostic techniques. Then at the end of the book, there is an appendix listing the various useful notations used in this book.

References of Chapter 1

[1] H. Holzler and H. Holzworth, *Pulstechnik*, (Springer-Verlag, Berlin, 1975)

[2] S. L. Shapiro, *Ultrashort Light Pulse, Pecosecond Techniquies and Applications*, (Springer-Verlag, Berlin, 1977)

[3] T. H. Martin et al., *6th IEEE Pulsed Power Conf.*, (1987) 225.

[4] J. Meyer-ter-vehn, Ed, *5th Int. Workshop on Atomic Physics for Ion Driven Fusion*, (Schliersee FRG 1990).

[5] G. H. Canavan et al., *Physics Today*, Nov. Issue, (1987), 48.

[6] M. Kristiansen et al., *Physics Today*, June. Issue, (1990), 36.

[7] *Aviation Week and Space Technology*, Aug. 22, Issue, (1988), 81.

[8] J. J Ramirez et al., *7th IEEE Pulsed Power Conf.*, (1989), 26.

[9] B. Henderson, *Aviation Week and Space Technology*, May 7, Issue, (1990), 88.

Chapter 2
ENERGY STORAGE

2-1. General Description

As indicated in Fig. 1-1, in a typical pulsed power facility, the required energy is commonly accumulated from the energy source at relatively low power levels and subsequently stored. Depending on the applications and requirement, the storage may be either in a capacitive form or an inductive form or a combination of the two. In the cases of the capacitive energy storage, the device usually consists of a number of high voltage capacitors connecting either in parallel or in series discharge configuration. The former is usually referred as a capacitor bank (Fig. 2-1) and the latter is called a Marx generator (Fig. 2-2). In both cases they are charged in parallel configuration. The capacitor banks can supply large amounts of current and usually are used as current source whereas the Marx generators can supply both high voltage and large current and therefore can be used as high power source.

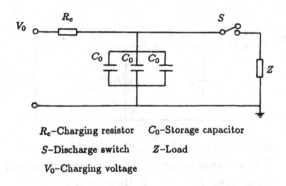

R_c–Charging resistor C_0–Storage capacitor

S–Discharge switch Z–Load

V_0–Charging voltage

Fig.2-1 A typical capacitor bank.

As for the inductive energy storage, instead of using capacitors, a magnetic inductor such as induction coil is generally employed to temporarily store the energy (Fig. 2-3). Unlike the capacitive case where the energy is directly transferred to the load by closing the switch, in this case the energy is first transferred from the capacitive storage to the inductive storage (coil in this case) and then transferred to the load. Therefore in the inductive storage system, in addition to the closing switch S_1, another switch S_2 (open switch) is required. The second switch S_2 serves the function to open the first loop in Fig. 2-3 at a proper time so that

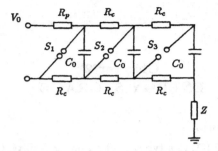

V_0–Charging voltage R_c–Charging resistor

R_p–Protecting resistor C_0–Energy storage capacitor

Z–Load S_n–nth switch

Fig.2-2 A 3-stage Marx generator.

the energy stored in L can be effectively transferred to the load by closing S_1. Synchronization of the two switches is critical in operating this type of system. Successful operation may significantly improve the energy transfer efficiency. We shall further discuss this aspect and other related aspects, respectively, in the following sections.

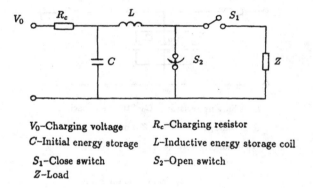

V_0–Charging voltage R_c–Charging resistor

C–Initial energy storage L–Inductive energy storage coil

S_1–Close switch S_2–Open switch

Z–Load

Fig.2-3 A typical inductive energy storage circuit.

2-2. Capacitor Bank

(a) Basic Principle

The circuit given in Fig. 2-1 is rather an ideal representation in which the inductance and resistance arising from various sources such as the capacitors, transmission lines, connecting leads, switch and the load, etc. have been neglected. In the actual situations these inductors and resistors cannot be generally ignored. When S closes the equivalent circuit of Fig. 2-1 it should be represented approximately by a RLC series circuit (Fig. 2-4), here C stands for the effective capacitance arising from all the C_o's.

Depending on the relative magnitudes of the R, L and C values, for $R < 2(L/C)^{1/2}$ (which is usually the case) the current i in the circuit may be written as

$$i = i_o e^{-\alpha t} \sin \omega t \qquad (2\text{-}1)$$

and the initial rate of current rise is

$$\left(\frac{di}{dt}\right)_m \approx \frac{V_o}{L} \qquad (2\text{-}2)$$

where $\alpha = R/2L$, $\omega = [1/(LC) - R^2/(4L^2)]^{1/2}$, $i_o = [(\omega^2 + \alpha^2)/\omega]CV_o$.

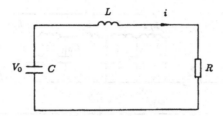

Fig.2-4 Equivalent circuit of Fig.2-1.

From Eq.(2-2), we can see that, regardless of what values for other parameters, the maximum rate of current increase is inversely proportional to L. That is, the smaller the L the faster the current rise time. Therefore, for fast pulse it is always desirable to keep the inductance of the circuit minimal. There are several ways that may reduce the inductance of the system, for instance, using low inductance capacitors, selecting proper dimensions for the transmission lines and leads, employing multichannel or multiple parallel switches and so on. Using multiple parallel switches (Fig. 2-5) has another advantage, the current going through each individual switch can be significantly reduced hence increasing the switch's life time. In doing so, however, some price will have to be paid, because synchronization of the parallel switches in this case becomes necessary, otherwise the system may not function properly. For example, in Fig. 2-5, if the first switch not function properly. For example, in Fig. 2-5, if the first switch s_1 closes prematurely, unless the transmission line between points a and b is sufficiently long, the potential at b will quickly rise to V_o as a result the switches s_2 and remaining switches may not close. One way to overcome this problem is to employ external trigger so that each of the switches is externally triggered.

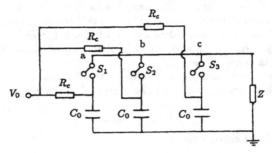

Fig.2-5 Capacitor bank using parallel switches.

As for achieving fast current risetime, one alternative to the above approach is to employ the so called transmission line network as shown in Fig. 2-6.

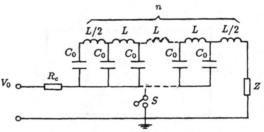

Fig.2-6 Transmission line equivalent capacitor bank circuit.

If the stage number n is sufficiently large, the network may be approximated as an uniform transmission line. The output from such a system can be nearly a step function with a maximum current of

$$i_m = \frac{V_o}{\sqrt{L/C_o} + R} \qquad (2\text{-}3)$$

and pulse width of

$$T = 2(n-1)\sqrt{LC_o} \qquad (2\text{-}4)$$

current rise time of

$$t_r \approx \frac{0.4T}{n} \qquad (2\text{-}5)$$

network impedance of

$$Z_o = \sqrt{L/C_o} \tag{2-6}$$

where R is the total resistance (including the load) of the network.

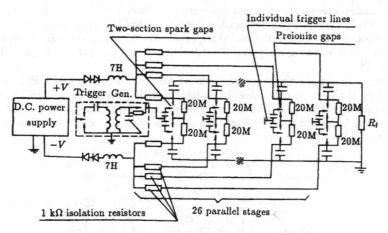

Fig.2-7 Bipolar charging, repetitive capacitor bank (26 stages)[1].

As such network is essentially a transmission line, multi-reflections of the current pulse at the ends may result if the output impedance does not match the network impedance Z_o. Proper care has to be taken when choosing the respective values of L, C_o and Z for such types of circuit.

(b) Bipolar Charging and Repetitive Systems

In order to obtain higher output voltage, capacitors banks are frequently charged in bipolar configurations in which half of the capacitors are positively charged and the other half are negatively charged. During discharge they form a series configuration hence yielding a voltage that is twice the original one. Bipolar charging system may either use single switch arrangement as that shown in Fig. 2-1 or use multiple switches arrangement similar to that in Fig. 2-5. For various reasons, however, the latter approach has been found to be more advantageous, particularly when the system is used in circumstances where continuous rep-rated operation is required. Because in the rep-rated operations, damage to the switch electrodes could be very severe if the total current is allowed to go through a single switch. Fig. 2-7 shows one of such systems which is designed for repetitive operations at 5 to 10 Hz[1]. This system consists of 52 0.75 μf capacitors. Half of them are positively charged to +30 kV and the other half are charged to −30 kV through a 7 H inductor. The total inductance of the circuit is 12 nH and the total energy stored is 50 kJ. There are 52 spark gap switches arranged in parallel configuration and all are externally triggered by a pulse generator of 250 kV, 50J capacity. The 20 MΩ resistor across each spark gap is designed to provide some discharge path so that an occasional switch malfunction such as a prefire or hangfire would only produce negligible effect on the output of the whole system. For example, if one switch malfunctions, its effect on the output voltage would be no more than 1% to the whole system. The system has been operated continuously for more than 50000 shots with no malfunction taking place. With similar arrangements, capacitor bank of 9.5 MJ capacity and 90 MA short circuit current has been built and operated[2].

Evaluation of the performance of the system may be done by using Eqs. (2-1) and (2-2), if the total resistance (including load) is known. In practice, however, such a quantity is not usually readily available, and especially the switch resistance is not time-independent. We shall discuss this aspect in more detail in a latter section.

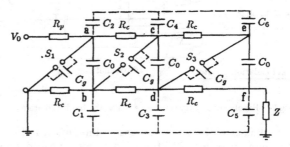

Fig.2-8 Marx generator circuit with stray capacitor included. C_g–capacitance due to switches; $C_1, C_2 \cdots C_n$ are stray capacitances and they are not necessarily all identical. The rest symbols may be identified from Fig.2-2.

2-3. Marx Generator

(a) Basic Principle

Analogous to the capacitor banks, the Marx generator also employs capacitors to store the energy. The main difference is that in this case when it is discharged all the capacitors are momentarily converted into a configuration of series connection. Therefore the Marx generator not only can serve as an energy storage but also can be used as a voltage amplifier. If there are N stages, in principle the output voltage should be N times the original voltage. In practice, however, due to various limitations the output voltage generally is considerably lower than NV_o. The circuit given in Fig. 2-2 is only an ideal representation in which the spark gap capacitance and the stray capacitance of the surroundings have been totally neglected. The real circuit should be modified in a way as that shown in Fig. 2-8. Let's use the portion a, b, c, d to make an analysis. When switch S_1 closes, the situation surrounding the switch S_2 may be approximately represented as that shown in Fig. 2-9. From Fig. 2-9, we can see that the potential at point c is:

$$V_c \approx V_o - V_o \frac{C_g}{C_g + C_3 + C_4} = V_o \frac{C_3 + C_4}{C_g + C_3 + C_4}$$

and

$$V_{cb} = V_c - V_b = V_o \left[1 + \frac{C_3 + C_4}{C_g + C_3 + C_4} \right] \qquad (2\text{-}7)$$

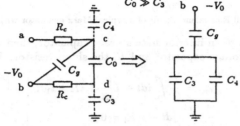

Fig.2-9 Approximate representation of the portion a, b, c, d in Fig.2-8.

From Eq.(2-7), we can see that only when $C_g \ll (C_3 + C_4)$ then $V_{cb} = 2V_o$, otherwise $V_{cb} < 2V_o$. In actual cases, usually $V_{cb} = 1.5V_o$. That means the switch S_2 will see a voltage considerably less than $2V_o$ when S_1 closes. By successive application of the formula (2-7) one can find the voltage across each of the remaining switches.

In the previous sections, the effects of the charging resistors R_c on the output voltage have been totally ignored. This is generally acceptable for high power pulse applications as in such cases the pulses encountered are usually considerably fast (mostly in the ns to μs range), the R_c effect on the output are therefore negligible. However, there may be situations that the pulse width encountered is much longer, e.g. in the hundreds of μs range, then such effect can't be totally ignored and proper care should be taken. As the discussion of such topics may be found in many H.V. text books, we will not discuss it in any details here.

(b) Circuit Analysis

The equivalent circuit of a Marx generator with load may be approximately written as that shown in Fig.2-10. In Fig.2-10, $C = C_o/N$ is the effective capacitance, V_m is the maximum voltage of the Marx generator. L and R are respectively the total inductance and resistance arising from all sources. C_l and R_l are respectively the load capacitance and resistance. V_l is the load voltage. Such circuit is typical in high power pulse applications. For example, when a Marx generator

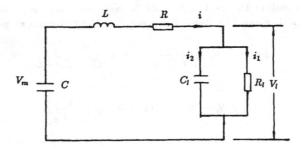

Fig.2-10 Equivalent circuit of a typical Marx generator with load.

is employed to charge an intermediate storage capacitor, then C_l would represent the intermediate storage capacitor and R_l the shunt resistor. From Fig. 2-10, we have:

$$\left.\begin{array}{l} V_m = \dfrac{1}{C}\int_0^t idt + L\dfrac{di}{dt} + iR + i_1 R_l \\[2mm] \dfrac{1}{C_l}\int_0^t i_2 dt - i_1 R_l = 0 \\[2mm] i_1 + i_2 - i = 0 \end{array}\right\} \qquad (2\text{-}8)$$

By solving Eqs.(2-8) one may get the voltage across the load as

$$V_l = V_m \alpha^2 \left\{ \frac{e^{\lambda_1 t}}{(\lambda_1 - \lambda_2)(\lambda_1 - \lambda_3)} + \frac{e^{\lambda_2 t}}{(\lambda_2 - \lambda_1)(\lambda_2 - \lambda_3)} \right. $$
$$\left. + \frac{e^{\lambda_3 t}}{(\lambda_3 - \lambda_1)(\lambda_3 - \lambda_2)} \right\} \qquad (2\text{-}9)$$

where λ_1, λ_2 and λ_3 are the three roots of the equation,

$$G^3 + (\frac{1}{\tau_1} + \frac{1}{\tau_2})G^2 + (\gamma^2 + \frac{1}{\tau_1 \tau_2})G + \frac{\beta^2}{\tau_2} = 0 \qquad (2\text{-}10)$$

where $\alpha = 1/\sqrt{LC_l}$, $\beta = 1/\sqrt{LC}$, $\gamma = \sqrt{(\alpha^2 + \beta^2)}$, $\tau_1 = L/R$, $\tau_2 = R_l C_l$, and the current going through the load may be obtained from

$$i_2 = C_l \frac{dV_l}{dt} \qquad \text{and} \qquad i_1 = \frac{V_l}{R_l} \qquad (2\text{-}11)$$

The expressions given in Eqs.(2-9) and (2-11) are too complex for analytic analysis, some simplification is obviously needed. For the cases $C_l = 0$, the circuit reduces into a typical $R_t LC$ series circuit with $R_t = R + R_l$. If further $R_t < 2(L/C)^{1/2}$, then the current and other relevant quantities may be obtained from Eqs. (2-1) and (2-2), provided the values of L, R and R_l are all known.

This, however, is generally not the case, because the inductance and resistance contributed from the switches are not constants and are strongly time-dependent. Unlike the capacitor banks in which the switches are in parallel configuration, inductance and resistance contributed from the switches are usually negligible and therefore the total inductance and resistance may be taken as constants. In the Marx generator all switches are effectively connected in series and contribution from the switches are significant hence the time-dependent effect cannot be generally ignored. In particular during the initial stage of closing, the switch's resistance may change several orders of magnitude. Therefore, in Eq.(2-8) assuming R and L are constants is obviously inaccurate. For more accurate evaluation of the current and other quantities one should take the time dependent effect into consideration.

In high power pulse applications, current or voltage pulses with fast risetime are frequently needed. From Eq. (2-2), we know that the risetime is roughly proportional to the inductance of the circuit, therefore making the circuit inductance small is one basic approach to achieve fast pulses. However such an approach is limited by the value of the inductance which in most case cannot be readily reduced. Another approach is to employ the so called peaking capacitor at the output end of the Marx generator. By means of this technique, voltage pulse of 600 kV with 1 ns rise-time has been achieved[3]. Let's use the circuit in Fig. 2-10 to illustrate how this can be done and what are the limitations when applying this method.

Assuming that in Fig. 2-10, $C_l = 0$, then the circuit may be treated as a typical RLC series circuit similar to that shown in Fig. 2-4. The current resulting from such circuit may be in oscillatory form or in aperiodic form depending on the relationship $R_t \neq 2\sqrt{L/C}$, here $R_t = R + R_l$ is the total resistance of the circuit. For $R_t < 2\sqrt{L/C}$, the current is in the form given by Eq.(2-1), except R is replaced by R_t and V_o by V_m. The current risetime may be approximated as

$$t_r \approx \frac{2.2}{w} \qquad (2\text{-}12)$$

where $w = (1/LC - R_t^2/4L^2)^{1/2}$ is the oscillatory frequency. If, on the other hand, $R_t > 2(L/C)^{1/2}$, then the current is aperiodic and may be expressed by the relationship:

$$i = \frac{\alpha^2 - \beta^2}{2\beta} CV_m e^{-\alpha t} \left(e^{-\beta t} - e^{\beta t} \right) \qquad (2\text{-}13)$$

where $\alpha = R_t/(2L)$, $\beta = [R_t^2/(4L^2) - 1/(LC)]^{1/2}$ and the current risetime may be approximated as

$$t_r \approx 2.3 \frac{L}{R_t} \qquad (2\text{-}14)$$

Now let's place a small capacitor $C_p (C_p \ll C)$ at the end of the Marx generator and a switch S before the load R_l, then the circuit would appear as that shown in Fig. 2-11. Before S closes, the situation is identical to that shown in Fig. 2-4, except C is replaced by C', here $C' = CC_p/(C + C_p)$ for $C >> C_p$ then $C' = C_p$. The current i_1 resulting from such a circuit again may be oscillatory or aperiodic depending on the circuit parameters. The current risetimes are identical to that given in equations (2-12) and (2-14) except replacing C by C' and R_t by R for the present case. Let the switch S close at a time when the capacitor C_p is charged to a maximum voltage, say V_m, then the loop on the right side may be momentarily considered to be a discharging circuit of capacitor C_p, as shown in Fig. 2-12.

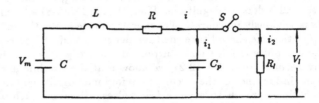

Fig.2-11 Marx generator with peaking capacitor C_p, before and after S closes.

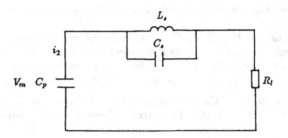

Fig.2-12 Momentary equivalent circuit of the loop on the right in Fig.2-11.

In Fig. 2-12, L_s and C_s are respectively the switch inductance and capacitance. For most practical cases C_s is very small and may be neglected, therefore the circuit may be again treated as a RLC series circuit and its current i_2 may be expressed by the same expressions given in Eqs.(2-1) or (2-13). The current risetime for the oscillatory case should be

$$t'_r \approx \frac{2.2}{\sqrt{1/(L_s C_p) - R_l^2/(4L_s^2)}} \tag{2-15}$$

and for the aperiodic case is

$$t'_r \approx \frac{2.3L_s}{R_l} \tag{2-16}$$

When the values of the various quantities given in the above expressions are properly chosen, the risetime t_r' of the current i_2 may be made significantly shorter than that given by Eqs.(2-12) and (2-14). As a result, the risetime of voltage pulse across the load may accordingly become very short while its falling part retains its slow characteristics. Such voltage pulse is important for many applications such as in the EMP generator where a long pulse with fast risetime is essential. Fig. 2-13 shows one typical pulse of this nature.

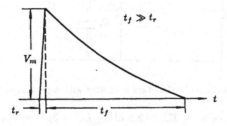

Fig.2-13 A typical EMP pulse.

Table 2-1 Specifications of the EMP Generator
designed by maxwell Co.

Output voltage	50 kV–450 kV
Risetime (10%–90%)	10 ns
Falling time (1/e)	800 ns
Rep-rate at 50 kV	1 Hz
Pre-pulse	< 5%
Output impedance	50 ohm
Insulation	SF_6

The physical explanation to such a technique may be understood qualitatively from the following argument. In Fig 2-11, when the switch S closes, discharging currents from both capacitors C and C_p are going through the load hence both currents i and i_2 contribute to the voltage drop across the load R_1. But the risetime of i_2 is much shorter than that of i, therefore the front portion of the voltage pulse is mainly contributed from i_2 whereas the current i which has a slow risetime and long pulse width is largely responsible for the falling part of the voltage pulse.

To apply such a technique effectively, some requirements have to be satisfied. From Eqs.(2-12) and (2-14), we can see that the risetime for the oscillatory case is generally longer than that for the aperiodic case. It is therefore necessary to make the loop on the right side in Fig. 2-11 to be aperiodic and the left one oscillatory. This in turn requires the values of all the circuit elements involved to be properly chosen, otherwise one may end up with some unexpected result. To see this, lets use a specific example to illustrate it quantitatively.

Fig. 2-14 is the equivalent circuit of a 50 ohm EMP generator designed by Maxwell Co. of the USA[4]. The designed specifications of the generator are given in Table 2-1.

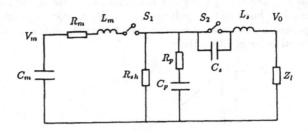

Fig.2-14 Equivalent circuit of a Marx generator with a peaking capacitor added [4].

In Fig. 2-14, $C_m = 6$ nf, $R_m = 2.5$ ohm, $L_m = 2\mu$H which are respectively the storage capacitor, circuit resistance and circuit inductance of the Marx generator, $R_{sh} = 800$ ohm is the Marx shunt resistance, $C_p = 1.2$ nf is the peaking capacitor, $R_p = 1.6$ ohm is the internal resistance of C_p, $C_s = 30$ pf and $L_s = 200$ nH are respectively the capacitance and inductance of switch S, $Z_l = 50$ ohm is the load impedance.

If there was no peaking capacitor present, the circuit may be taken as an aperiodic RLC series circuit. From equation (2-14) we know that the risetime of the current may be approximated as:

$$t_r \approx 2.3L/R_t \approx 90 \quad \text{ns}$$

However, when a peaking capacitor of 1.2 nf is added to the circuit the loop on the right side becomes a fast discharge circuit and produces a current with faster risetime t_r' after the switch S_2 is closed. From equation (2-16) and using the data given in Fig. 2-14, we can obtain an approximate value of t_r' as

$$t_r' \approx 2.3L_s/(R_p + Z_l) \approx 9 \quad \text{ns}$$

which is only one tenth of the value when there was no peaking capacitor employed. Fig. 2-15 is the calculated result using the data of Fig. 2-14. From these results we can see that by employing a peaking capacitor the risetime of the voltage pulse can be greatly improved. However, to achieve this objective proper care has to be taken when choosing the various elements in the circuit. For example, if in Fig. 2-14 one uses a 20 ohm impedance as the load instead 50 ohm and keeping the rest of the elements unchanged, one may wonder what would be the outcome. By a simple calculation one can readily find out that with a 20 ohm load, the loop on the right side in Fig. 2-14 is no longer aperiodic but becomes oscillatory hence the current risetime t'_r has to be expressed by Eq.(2-15) instead of Eq.(2-16). From Eq.(2-15) we find that

$$t'_r \approx 60 \quad \text{ns}$$

which is about 6 times slower than that when a 50 ohm load was employed. From this example we can clearly see the necessity of choosing proper values for the circuit elements in question.

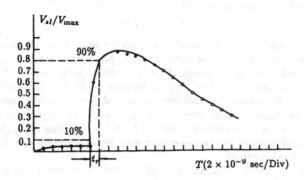

Fig.2-15 Calculated voltage pulse using data given in Fig.2-14.

(c) Evaluation of Time-dependent Effects

In the previous sections, we have treated the inductance and resistance of the switches as constants. In reality they are all time-dependent, therefore the first Eq. given in (2-8) should be modified as:

$$V_m = \frac{1}{C} \int_0^t i\,dt + L\frac{di}{dt} + i\frac{dL}{dt} + Ri \qquad (2\text{-}17)$$

In Eq.(2-17) L and R are the total inductance and resistance of the Marx circuit and both of them consists of two parts. One steady part L_o and R_o arising from the lump capacitors, connectors of the circuit which usually have fixed values and another time-dependent part L_t and R_t mostly due to the switches that vary with time, i.e.

$$\left. \begin{array}{l} L = L_o + L_t(t) \\ R = R_o + R_t(t) \end{array} \right\} \qquad (2\text{-}18)$$

If both the steady parts and the time dependent parts are known explicitly, then in principle Eq.(2-17) may be solved analytically. In the following, we shall illustrate respectively how each of these quantities may be evaluated.

(I) Steady parts: $L_o = L_1 + L_2$; $R_o = R_1 + R_2$

Here L_1 and L_2 are the respective inductance of the capacitors and connectors. The inductance due to the capacitors is generally given by $L_1 = nl_o$, where n

is the number of capacitors employed and l_o is the inductance associated with each individual capacitor which can be usually obtained from the manufacturers manual. For the inductance due to connectors and wires:

$$L_2 = 2l[\ln(2l/r) - 1] \quad \mu H \tag{2-19}$$

where l is the length (cm) of the connectors and r is the radius (cm).

Let R_1 and R_2 be the respective resistance of the connectors and contact resistance, for R_1, it is given by

$$R_1 = \rho(l/A) \tag{2-20}$$

and the R_2 expression is considerably complex which depends on many factors such as the contact area, surface nature and applied pressure etc., in many cases its estimated value is used. Using the expressions given above, all inductances and resistances in the steady part except R_2 may be readily calculated. However, results obtained from such calculations are not always accurate enough owing to various practical reasons. One alternative approach is to employ an experimental method to determine the effective values of L and R. For example, the inductance and resistance of a given Marx generator may be experimentally determined by using the following method.

Assuming the effective capacitance of a given Marx generator is $C_m = 20$ nf and the short circuit current is given in Fig. 2-16. From the above result we know that the period of the oscillation is $T = 3.45$ μs, since $T = 2\pi(L_m C_m)^{1/2}$ therefore we have $L_m = 15.07\mu$H.

Next we measure the current peak values from Fig. 2-16 and find the ratios between two successive peaks as that shown in Table 2-2.

Table 2-2 Ratios of current peaks

i_1/i_3	i_3/i_5	i_5/i_7	i_7/i_9	Average
1.383	1.424	1.375	1.411	1.428

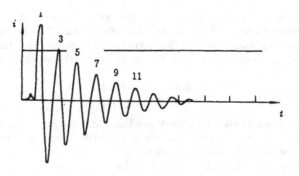

Fig.2-16 Short circuit current of a given Marx generator.

Finally from the relationship

$$\frac{i_n}{i_{n+1}} = \frac{i_o \exp(-\alpha t)}{i_o \exp[\alpha(-t - T)]} = \exp(\alpha T)$$

we get $\alpha = R/2L_m = 0.103 \times 10^{-6}$ henry, $R = 2L_m\alpha = 3.09$ ohm

(II) Time-dependent parts

For the determination of the time-dependent parts, L_t and R_t, there are no simple formulas as that given in Eqs.(2-19) and (2-20) may be used. Evaluation and analysis of these quantities have to rely on tedious procedures and quite often these procedures are not accurate. In a latter chapter, when we discuss the switches, we shall present a full discussion on the subject in more details. At this moment, let's simply introduce the expressions without further elaboration.

If gas spark gap switches are employed, the inductance of the plasma channels may be roughly expressed by[5]:

$$L_t \approx l\frac{\mu_o}{2\pi}\ln[r_c/r_s(t)] \qquad (2\text{-}21)$$

where l is the length of the gap; r_c is the radius of the current return path; $r_s(t)$ is the time-dependent radius of the plasma channel which may be determined by means of pulsed laser interferogram.

Alternatively, L_t may also be evaluated by using the empirical formula[6]:

$$L_t \approx kV^n t^m \qquad (2\text{-}22)$$

where V is the applied voltage to the switch; t is time; k, n and m are constants for a given medium. For water dielectric: $k = 5.6 \times 10^{-10}$; $n = 1.1$; $m = 0.67$.

For the resistance R_t, it may be either expressed by[7]

$$R_t = \left[\frac{2LPd^2}{V_m Ca(t/2 - (1/4\omega)\sin 2\omega t)}\right]^{1/2} \qquad (2\text{-}23)$$

where $\omega = 1/(LC)^{1/2}$; P is the pressure; d is the gap separation; a is a constant for a given gap, or by the empirical formula[7-9]:

$$R_t = k' P^l E^m t^n e^{-(t/\tau)} \qquad (2\text{-}24)$$

where k' is a constant for a given dielectric; P is the gap pressure; E is the applied E-field; l, m, n and τ are all constants for a given medium. However, these constants given in Eq.(2-24) differ considerably between different models, for example, from reference [8]:

$$R_t = \frac{88R^{2/3}}{E^{4/3}}\left(\frac{\rho}{\rho_o}\right)^{1/2} t^{-1}e^{-t/\tau} \qquad (2\text{-}25)$$

where $\tau = [(L/R) + (88/R^{1/3}E^{4/3})(\rho/\rho_o)^{1/2} \times 10^{-9}]$; R: load resistance; ρ_o: gas density at 1 atm; ρ: gas density in the gap; L: total inductance; E: the applied field in 10 kV/cm. But from reference [9] for N_2 gas:

$$R_t \approx 2P^{3/2}E^{-3}Z_o^{-1}t^{-3} \times 10^4 \qquad (2\text{-}26)$$

where Z_o is the impedance of the transmission line.

Further from reference [5],

$$R_t = kF_{sb}^{3/5}\pi^{-2/5}[I(t)]^{-6/5}P^{3/5}t^{-4/5} \qquad (2\text{-}27)$$

where F_{sb} = charge voltage/self-break voltage; $I(t)$ = current going through the switch.

By substituting quantities given in expressions through (2-19) to (2-27) into Eq.(2-17), in principle the current i and other relevant quantities may be obtained. In practice, however, Eq.(2-17) can't be readily solved analytically, unless significant simplifications are made. The general approach is to rely on a numerical solution.

(d) Aspects Concerning Design and Application

In high power pulse field, the demand for Marx generators is great. It has to be capable of holding high voltage, delivering large current, reliable, long life and finally it is also required to be compact so that a group of Marx generators can be concurrently employed without using large space. At present, there is no unique approach that can satisfy all these demands. There is always some conflict present when one wishes to achieve different goals at a time. For example, if one uses a large number of stages to increase the voltage, then the reliability will become poor. Similarly if large capacitors are employed to increase the current, then the compactness and the switches lifespan will suffer. The practical approach is to tackle the problems on an individual basis such that the most important requirement is satisfied and leave the less important ones aside or else make a compromise between them.

Bipolar charging technique appears to be one practical approach as it provides a way to use a large number of stages while keeping the volume relatively small. As for the reliability problem arising from too many stages, one solution is to choose $\alpha = V_{sb}/V_{ch}$ and $\beta = V_{tr}/V_{sb}$ both having high values and use powerful trigger pulse to trigger each of the switches externally as that shown in Fig. 2-17, here V_{sb} = self-break voltage, V_{ch} = charging voltage, and V_{tr} = trigger pulse voltage. In this particular example, the trigger pulses are sent through series configuration, hence at the end stage the β value may become considerably weaker. To maintain a high value of β, the trigger pulses have to be sent through parallel configuration as that shown in Fig. 2-7. With such an approach, however, another problem may arise, i.e. one would need a powerful pulse generator to provide enough current for each of the parallel triggers as there is always some current going

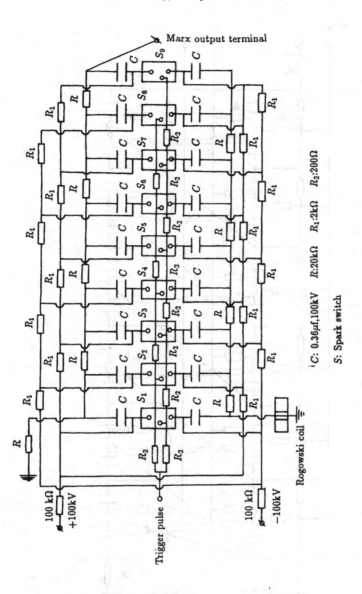

Fig.2-17 Bipolar charged Marx generator with series trigger configuration.

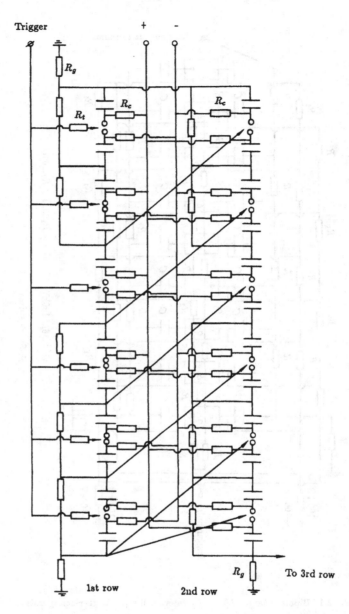

Fig.2-18 Bipolar charged Marx generator with mixed trigger scheme.

though the trigger electrodes. To lessen such a defect, one solution is to employ internal-external mixed trigger scheme[10]. In this scheme, only the very first row of switches is externally triggered, and the remaining switches are triggered internally by overvoltage pulses coming from the previous stages as shown in Fig. 2-18. With this scheme, a pulse generator of moderate energy would suffice to trigger the Marx generator effectively. The specifications of one of such designs reported in Ref.[10] are given in Table 2-3.

Table 2-3 Specifications of one of the Marx generators using mixed trigger scheme

$V_o = \pm 100$ kV	$L = 11.3$ μH
$V_m = 6$ MV	$R = 1.8$ ohm
$C_o = 1.3$ μF	$R_s = 1450$ ohm
$C = 21.7$ nF	$\Delta t = 250$ ns
$E_t = 390$ kJ	$\Delta \sigma = 7$ ns
$V_{tr} = 500$ kV	

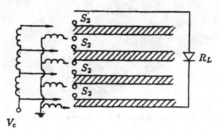

Fig.2-19 Basic structure of a 4-stage strip line Marx generator. V_c– charging voltage; S–closing switch; R_L–load.

In addition, there are many other types of Marx generators which are usually designed to meet specific requirements. One of them is the strip line Marx generator which is designed to produce short voltage pulses[11]. The Marx generator of this type uses strip transmission lines instead of capacitors as the energy storage. The basic principle for producing high power pulses is similar to that for the conventional Marx generators, namely the strip transmission lines are charged in parallel configuration and discharged in series configuration. Fig. 2-19 shows the basic structure of a typical 4 stage strip line Marx generator. As each transmission line in principle is capable of producing voltage pulses of step risetime, a series connection of them could serve as a fast pulses generator is conceivable. The capability of producing fast pulses is in fact the main merit that makes the strip line Marx generator attractive. With reasonable cost and effort, one can readily build a strip line Marx generator capable of producing fast pulses that normally cannot be achieved with the conventional ones. The main drawback of the strip line Marx generator is the relatively large size. Because of its geometrical configuration, it is difficult to make it as compact as that of a conventional one having the same capacity. Fig. 2-20 is a photo of a 50 stage, 400 kV strip line Marx generator built by The Chinese Academy of Sciences. Table 2-4 shows the specifications of this generator along with another one built by the same institute. From these figures we can see that at 1000 kV voltage, 40 ns pulse width is considerably fast.

Table 2-4 Specifications of two models of strip line Marx generator

Specifications	model I	model II
Number of stages	50	100
Pulse peak voltage	400 kV	1000 kV
Pulse peak current	4 kA	4 kA
Pulse width	40 ns	40 ns
Source impedance	125 ohm	250 ohm

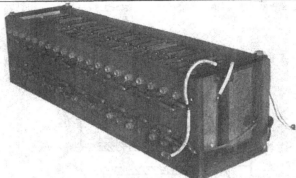

Fig.2-20 A 50 stage, 400kV strip line Marx generator built by the Chinese Academy of Sciences [11].

Another type of Marx generator capable of producing fast pulses is called the "compartment Marx generator". Its basic structure is shown in Fig. 2-21. It consists of a number of identical compartments which can be readily assembled together or dismantled by means of the plug-in unit attached to it. This feature enables the user to freely adjust the number of stages needed. Each stage contains a number of ceramic capacitors in parallel configuration such that the inductance is reduced. Owing to the small capacity of the ceramic capacitors, such a Marx generator usually cannot supply a large amount of current as the conventional ones do. However, it does provide the user with great convenience when the output voltage need to be adjusted. Hewlett-Packard Co. of the USA has developed several models of this type Marx generator[12]. The basic characteristics of the model 43733A is shown in Table 2-5.

Table 2-5 Specifications of model 43733A

Number of compartments	12
Charging voltage	25 kV
Output voltage	300 kV
Output current	5 kA
Pulse width	30 ns
Voltage efficiency	50 %

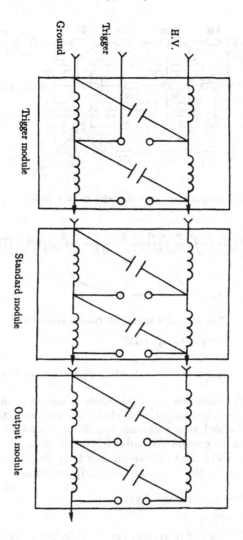

Fig.2-21 The basic structure of a compartment Marx generator[12].

In certain applications, such as in the developments of high power lasers, long pulses (μs) with fast rise and falling times (ns) are frequently needed. The Marx generator shown in Fig. 2-22 has demonstrated the capability of meeting such

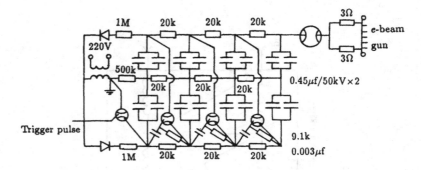

Fig.2-22 Bipolar charged Marx generator with crowbar at the output terminal.

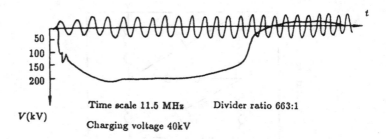

Time scale 11.5 MHz   Divider ratio 663:1

$V(\text{kV})$

Charging voltage 40kV

Fig.2-23 Output voltage from the Marx generator shown in Fig.2-22.

requirements[11]. The main feature of this Marx generator is that: (1) the capacitors in the storages are all arranged in parallel configuration so that the inductance is kept minimal, (2) a high power vacuum switch with fast switching time was employed as a crowbar to control the pulse width. Results obtained from such a generator is shown in Fig. 2-23. It can be seen from the figure that during the 1 μs or so pulsing duration the voltage stays nearly constant.

2-4. Inductive Energy Storage Systems

(a) General Description

Up to today, the capacitive system described in the preceding sections is still the most widely employed device for energy storage in the high power pulse applications. Other devices such as the inductive energy storage system has not gained similar popularity as that of the capacitive one. However, in terms of energy density, the capacitive energy storage system is not so efficient as compared to the inductive ones. The capacitive energy storage system has some fundamental constraints imposed by its physical characteristics and the relatively large size.

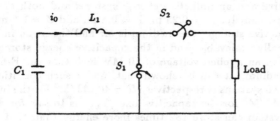

Fig.2-24 A typical inductive energy storage circuit.

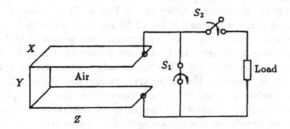

Fig.2-25(a) Inductive energy storage system.

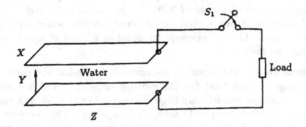

Fig.2-25(b) Capacitive energy storage system.

Fig. 2-24 is a typical circuit of inductive energy storage system. During the operation, in the beginning switch S_1 is closed and switch S_2 is kept open. When the current i_o reaches its maximum value switch S_1 opens and at the same time closes switch S_2, so that the energy stored in L_1 is transferred to the load. If the rate of resistance change in S_1 is rapid enough, transfer of the magnetic energy from L_1 to the load can be made very efficient. Intrinsically the energy density of a magnetic field can be made much greater than that of an electrostatic field, hence allowing the magnetic energy storage to be made significantly smaller in size. In the case of capacitive energy systems, numerous capacitors in parallel configurations would be required to achieve the same goal. For comparison, Fig. 2-25 shows respectively a capacitive and an inductive energy systems and both of them have the same dimensions, namely: $X = 0.25$ m, $Y = 0.4$ m, and $Z = 1.5$ m. Let's assume that in the inductive storage, the medium is air and the mean induction is 10 T which is practically achievable, and in the capacitive energy storage, the medium is water and with an applied voltage of 10 MV such that the E-field is tolerable to the water medium. It can be shown that, under such conditions, the energy density in the two systems is respectively $U = 40$ MJ/M^3 for the inductive system and $U = 0.22$ MJ/M^3 for the capacitive one. That is to say, for a given volume, the inductive system can store 182 times more energy than that of a capacitive one. In fact, this figure could be made as high as 10^3 if the optimal conditions are utilized. Furthermore, the powers that can be delivered to the load by the two systems are also significantly different. In Fig. 2-25(a), when S_1 opens and S_2 closes at the same time, a current pulse of 10 ns duration will be delivered to a matched load which in turn will produce a maximum power of 600 TW, whereas in Fig. 2-25(b), when S_1 closes, the maximum power that may be delivered to a matched load is no more then 0.4 TW that is only a small fraction of the former. This is because, in addition to the respective energy, density is different, the E-M wave propagation velocity in water is much slower than that in air, hence the pulse duration is much longer (83 ns) than the 10 ns of the inductive case. From this illustration, we can clearly see the advantages of the inductive systems. It not only can provide higher energy density but can also deliver higher power. For example, using the inductive energy storage in conjunction with a capacitors bank of 38 kV charging voltage, an output power of 0.26 TW has been obtained[13]. However, to utilize such advantages is not a straight-forward matter, some stringent requirements have to be met. One of them is to develop a kind of open switch that is capable of doing the task.

The requirements for the opening switches in an inductive system is considerably more stringent than that for the closing switches. The opening switch is required to be capable of conducting high current when it is closed and interrupting the current completely instantly when it is opened. This requires the switch to have high impedance and a high rate of impedance increase. Furthermore it is also required to hold off high voltages and have long conduction time and short recovery time. At present, it is not possible to design a switch that can simultaneously satisfy all these requirements, because the performance of the switch is not only determined by the switch's characteristics but also strongly coupled to the circuit parameters. The general approach is to first establish some basic rules basing on some approximate model and then using these rules as a guide to finally determine the switch parameters from the available empirical data. We shall leave the discussion on how to establish such rules as well as the selection of an opening switch in a latter section. Let's first look at the basic circuits that may be employed in the inductive energy systems.

In addition to the circuit shown in Fig. 2-24, there are a number of other types of circuits that may be employed to transfer the inductive energy to the load. They

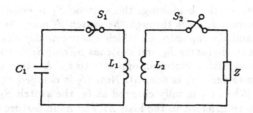

Fig.2-26 Transformer type inductive circuit.

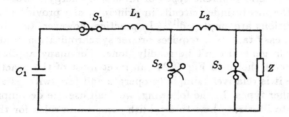

Fig.2-27 Meatgrinder type inductive circuit.

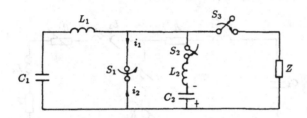

Fig.2-28 Counterpulse inductive circuit.

include the transformer type (Fig. 2-26), the meatgrinder type (Fig. 2-27) and the counterpulse type (Fig. 2-28). For the transformer type, during operation, the primary inductor L_1 is first energized and with the secondary coil kept open. After the capacitor C_1 has been discharged, the switch S_2 is then closed and meanwhile the switch S_1 in the primary is opened, hence the energy in L_1 is transferred to the load Z. In the meatgrinder circuit, during operation, the inductors L_1 and L_2 are first energized with the switches S_1 and S_3 closed and S_2 is kept open. After the capacitor C_1 has been discharged, S_3 is opened so that a current is established in the load Z. Following that the switche S_2 is closed and the switch S_1 is opened, hence the so called meatgrinding process takes place and the energy stored in the inductors is thus transferred to the load Z. The operation in the counterpulse circuit is as follows: at the beginning, the switche S_1 is closed, S_2 and S_3 are kept open so the current i_1 flows through the switch S_1 only. At a proper time t_1 switch S_2 is closed and the capacitor C_2 which was previously charged to V_2 is now discharged through the inductor L_2 and switches S_2 and S_1. Hence a current i_2 is going through S_1 and with a direction opposite to i_1. At time t_2 when the total current in S_1 becomes zero, S_1 is opened. Now C_2 is charged again but with the polarity reversed. When C_2 is fully charged at t_3, the switch S_3 is closed, hence the stored energy is transferred to the load Z. The main feature of this technique is to artificially create a zero current in S_1 so that it does not have to stand the inductive voltage of high magnitude.

Among the several different types of inductive energy storage circuits mentioned above, the meatgrinder circuit, in principle, can provide higher efficiency if optimal conditions are employed[14]. In practice, however, to achieve such efficiency is not an easy task as it requires proper synchronization of all the switches involved which in general can't be readily done. For many applications the decompression circuit shown in Fig. 2-24 can meet most of the practical purposes. This is because it is relatively simple to operate and few switches are involved as compared to other types. In the following, we shall use the decompression circuit as an example to illustrate how to establish some basic rules for the selection of an open switch in an inductive energy storage circuit. (The decompression circuit is so named because upon closing the switch S_2 in Fig. 2-24, the magnetic flux stored in L_1 expands from L_1 to the load).

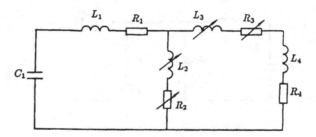

Fig.2-29 Equivalent circuit of that shown in Fig.2-24.

(b) Circuit Analysis

Among the several different types of inductive energy circuits, the decompression circuit shown in Fig. 2-24 is the simplest of all. However, for circuit analysis, the simplest one is not really so simple. Fig. 2-29 is the actual equivalent circuit of Fig. 2-24, when all switches are closed, i.e. when S_2 closes, yet S_1 has not completely open. In the figure, R_1 is due to the circuit conductors etc., L_2 and R_2 are due to the open switch S_1 and both are time dependent, L_3 and R_3 are due to the switch S_2 are also time dependent, L_4 and R_4 are due to the load. To analyze the circuit as such, one would have to use a numerical method and an analytical solution is obviously not possible. In order to get some analytical solutions, significant simplification of the circuit is necessary. We therefore, in the following discussion, shall assume that the circuit shown in Fig. 2-24 may be approximated on the basis of the following assumptions:

(I) all the quantities R_1, L_2, L_3 and R_3 in Fig. 2-29 are negligible.

(II) prior to S_2 closing, the current going through L_1 reaches its maximum value and the capacitor C_1 is completely discharged at $t = 0$.

(III) when S_1 closes, its resistance is zero but the resistance of S_1 is a function of time before S_1 is completely open.

With such approximations, when S_2 closes and S_1 has not completely opened the equivalent circuit of Fig. 2-24 may be represented as that shown in Fig. 2-30.

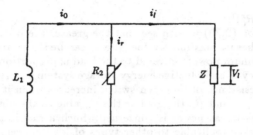

Fig.2-30 Simplified equivalent circuit of Fig.2-24.

From Fig. 2-30 , we have:

$$\left.\begin{array}{l} i_o - i_r - i_l = 0 \\ L_1 \dfrac{di_o}{dt} + i_r R_2(t) = 0 \\ i_r R_2(t) - V_l = 0 \end{array}\right\} \qquad (2\text{-}28)$$

To make the Eqs. given in (2-28) solvable analytically, the time function $R_2(t)$ has to be known. Here we make three different approximations:

(I) $R_2(t)$ is a linear function of time, i.e. $R_2 = mt$ (m is a constant).

(II) $R_2(t)$ is a step function of t, i.e. at $t = 0$, $R_2 = 0$, and at $t > 0$, $R_2 = R_f$ =constant.

(III) $R_2(t)$ is a step function, but has the property at $t = 0$, $R_2 = 0$, and at $t > 0$, $R_2 = \infty$.

In the following, we shall respectively discuss the outcomes of these three cases basing on Eqs. (2-28).

(I) For $R_2 = mt$, if the load is a pure resistor R_l, then in Eqs.(2-28) we can replace V_l by $i_l R_l$ and using the boundary condition at $t = 0$, $i_o = I_o$, we can get the current going through the load R_l, as

$$i_l = I_o \frac{mt}{mt + R_l} \exp[(R_l^2/mL_1)\ln((mt + R_l)/R_l) - (R_l/L_1)t] \qquad (2\text{-}29)$$

It can be shown that at

$$t = \tau_1 = \sqrt{L_1/m} \tag{2-30}$$

i_l is maximum and may be expressed by

$$(i_l)_m = I_o \frac{(mL_1)^{1/2}}{(mL_1)^{1/2} + R_l} \exp[(R_l^2/mL_1)$$
$$\times \ln\{1 + (mL_1)^{1/2}/R_l\} - (R_l/(mL_1)^{1/2}] \tag{2-31}$$

Hence the maximum power P_m that can be delivered to the load R_l is

$$P_m = (i_l)_m^2 R_l = \frac{I_o^2 R_l a^2}{(1+a)^2} \exp[(-2/a) + (2/a^2)\ln(1+a)] \tag{2-32}$$

where $a = (mL_1/R_l^2)^{1/2}$.

From expression (2-30) we can see that the greater the m value is, the faster the current i_l reaches its maximum value. It also can be shown from expression (2-32), that the maximum power delivered to the load is proportional to m. In other words, the efficiency of an inductive energy storage system is very much dependent on how fast the resistance of the open switch increases when it is opened. For a circuit of given L_1, I_o and R_l, the greater the m value is, the higher the efficiency can be achieved as far as power is concerned. Such a rule is not only valid for resistive load, it also applicable to other types of load as well. For example, if the load is a pure capacitor C_l, then in Eqs. (2-28), we simply replace V_l by $(\int_0^t i_l dt)/C_l$ and follow the same procedure as that outlined above, we may obtain the maximum power that can be delivered to a capacitive load of C_l as

$$P_m' = \sqrt{\frac{L_1}{C_l}} \frac{I_o^2 n!^2 2^{2n}}{(2n-1)^2} (\omega t)^{-2n+1} J_n(\omega t)[J_n(\omega t) - \omega t J_{n+1}(\omega t)] \tag{2-33}$$

where $n = (1/2)(1 + 1/(C_l m))$, $\omega = 1/(L_1 C_l)^{1/2}$ and $J_n =$ Bessel function of order n. Though the expression given in (2-33) is considerably complex, by numerical evaluation[15], it can be shown that the maximum power P_m' is again proportional to the value of m, i.e. the faster the switch S_1 opens, the greater the power will be delivered to the load C_l.

Finally, if the load is an inductor L_l, then in Eqs.(2-28) by replacing V_l with $L_l(di_l/dt)$, one can again obtain the maximum power as

$$P_m'' = 0.33 I_o^2 (mL_l)^{1/2} \frac{L_1}{(L_1 + L_l)^{3/2}} \tag{2-34}$$

Once again, it can be seen from (2-34) that the maximum power increases with $m^{1/2}$.

From expressions (2-32), (2-33) and (2-34) it can be seen that in all cases, the maximum power that can be delivered to the load is proportional to the current square I_o^2. Based on these analyses, it is clear that for an inductive energy storage circuit, whether the load is resistive, capacitive or inductive the most important factors that determine the efficiency of the entire circuit are the two parameters m and I_o. The greater these values are, the more efficient is the transfer of energy from the primary energy source to the load. As for the current I_o, if the resistance of the primary circuit can be minimized near zero, then I_o will be roughly proportional to $V_o(C_1/L_1)^{1/2}$, so it may be increased either by employing large capacitor C_1 or by higher charging voltage V_o or by both. Selection of those parameters in most cases are straightforward and is largely determined by economic factors rather than by technological reasons. However, selection of the parameters for an open switch having both high m value and capabilities meeting the other requirements is a rather complex matter. Firstly, at present, there is no well-developed technology that can be relied on to design such a switch. Secondly the performance of the open switch, as can be seen from the preceding analysis, is closely coupled to the circuit parameters. Therefore design of the open switch, in many cases, has to largely rely on empirical data with trial and error approach. Quite often, such approaches fail to give satisfactory results. This is one of the basic reasons why the inductive energy storage system has not been employed as widely as the capacitive one, though the former has apparent advantages.

(II) The above analysis only deals with the cases where the resistance of the open switch increases linearly with time. In some other cases, the switch resistance may be approximately represented by a step function of time with a finite final value R_f. For example, the resistance

$$R_2 = R_f(1 - e^{-t/\tau}) \tag{2-35}$$

may be approximated by such a step function, if the time constant τ is sufficiently small. Under such circumstances, analytic solution from Eqs. (2-28) is still possible. With the boundary conditions: at $t = 0$, $i_l = 0$ and $i_l(L_1 + L_l) = I_o L_1$, it can be shown that for the case of inductive load L_l, the load current may be expressed by

$$i_l = \frac{I_o L_1}{L_1 + L_l}(1 - e^{-\gamma t}) \tag{2-36}$$

where $\gamma = R_f(L_1 + L_l)/(L_1 L_l)$.

From expression (2-36), it can be seen that the power can be made significantly large, if the initial current I_o and the final resistance R_f are made large. This is similar to the conclusion drawn from the previous analysis.

(III) Finally, lets assume the resistance of the open switch undergoes a step rise to $R_f = \infty$, i.e. from $R_2 = 0$ to $R_2 = \infty$ stepwise. If the load is inductive and having a value of L_1, then from the law of conservation of magnetic flux, one may obtain the relation:

$$i_l = \frac{I_o L_1}{L_1 + L_l} \tag{2-37}$$

Now let's define the inductive efficiency of the circuit to be the ratio of the energy E_1 delivered to the load to the initial stored energy E_o, i.e,

$$\eta = \frac{E_l}{E_o} = \frac{i_l^2 L_l}{I_o^2 L_l} \tag{2-38}$$

Put Eq.(2-37) into Eq.(2-38) we have

$$\eta = \frac{L_1 L_l}{(L_1 + L_l)^2} = \frac{\alpha}{(1 + \alpha)^2} \tag{2-39}$$

where $\alpha = L_1/L_l$.

Similarly, we can show that for the transformer type circuit given in Fig. 2-26, its inductive efficiency is

$$\eta' = \frac{k^2 \beta}{(1 + \beta)^2} \tag{2-40}$$

where $\beta = L_2/L_l$, k is the coupling coefficient and $k \leq 1$.

For the meatgrinder circuit shown in Fig. 2-27, its inductive efficiency is

$$\eta'' = \frac{[\alpha + \beta + 2k(\alpha\beta)^{1/2}]^2 [1 + \beta + k(\alpha\beta)^{1/2}]^2}{[\alpha + \beta + 2k(\alpha\beta)^{1/2}](1 + \beta)^2 [1 + \alpha + \beta + 2k(\alpha\beta)^{1/2}]^2} \tag{2-41}$$

Under optimal conditions, it can be shown that among the three circuits mentioned above, the meatgrinder circuit can provide the highest efficiency and the transformer type circuit is the least efficient one[19].

From the above analysis, we can see that the key element in an inductive energy storage circuit is the open switches. In the following, we shall briefly review the open switch developments and some of the basic characteristics.

(c) Open Switches and their Characteristics

At present, there are about a dozen or so open switches of different designs in existence. Their capabilities and characteristics differ widely in terms of the two key parameters, namely the rate of resistance rise m and the initial maximum current I_o going through the switch. Schoenbach et al.[15] have compiled a summary of the presently available open switches and plotted in terms of the two parameters m and I_o. The result is reproduced in Fig.2-31. From the figure, we can see that as far as m is concerned, the diffuse discharge switch is obviously the best. However, in terms of the combined I_o and m capabilities, the three switches at the top namely the Fuse, Plasma Erosion and Explosive, have the best combinations. This is one of the reasons that these three switches have recently received the most attention from researchers in the field of high power pulse technology. Though the three switches have almost similar capabilities, the technology in making and operating them differs significantly. For example, utilization of the plasma erosion switch involves far more advanced technologies than that for the other two. It not

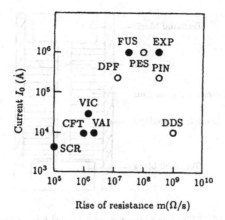

Fig.2-31 Current I_0 versus rise of resistance m, for high power opening switches. PES: Plasma erosion switch; EXP: Explosive; FUS: Fuse; PIN: Plasma instabilities; DPF: Dense plasma focus; DDS: Diffuse discharge switches; VAI: Vacuum switch; CFT: Crossed field tube; SCR: Silicon contr. rectifier; VIC: Vacuum interrupter(counterpulsed)[15].

only requires the sophisticated plasma gun but also needs a vacuum environment to operate in it[16−18], whereas in the case of fuse or explosive switch, the structure of the switch is relatively simple and easy to make, furthermore they can be operated in almost any kind of environments. For these practical reasons, the fuse and explosive switches have been considered to be the most useful open switches in the high power pulse community, in particular in the applications of inductive energy storage. In the following, we shall therefore concentrate on the discussion of these two types of switches only, for other types of open switches we suggest the readers refer to Refs.[15] and [20] where comprehensive discussions on the subjects are given.

Fuses employed in pulsed inductive energy storage system are operated on the same basis as that for any conventional fuse, namely they are operated as a consequence of rapid ohmic heating by the current which they carry, resulting in their melting and subsequent vaporization that causes their resistance to increase drastically and finally cut off the current. Their function in the pulsed power applications however is unlike that in the conventional cases, where fuses are used mostly for protection purpose, whereas in this case they are used mainly to enhance the output power by means of interrupting the current in the storage inductor thus leading to the formation of a fast rising current in the load. The interruption process can be very fast (50 ns) if the fuse material, dimensions and the surrounding medium are properly chosen[21].

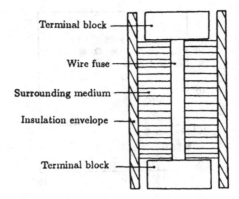

Terminal block

Wire fuse

Surrounding medium

Insulation envelope

Terminal block

Fig.2-32 The basic structure of a typical fuse switch.

Thin wires or foils embedded in some quenching medium have been commonly used to fabricate fuse switches. The basic structure of a typical fuse switch is shown in Fig. 2-32. The characteristics of a fuse up to the time of vaporization is largely determined by the dimensions and the material properties of the conductor. Prediction of such behavior is usually possible. After vaporization, however, the behavior is rather complex as a result of interaction between the metal vapor and the surrounding medium. Such interaction, depending on the circumstances, may lead to various consequences, such as expansion or contraction of the vapor channel, production of electrons by ionization or removal of electrons by attachment. All these processes can significantly affect the performance of the fuse. For example, when the vapor channel expands, the gas density decreases and an arc may occur and as a result the voltage across the fuse may be reduced drastically. Another mechanism which may lower the voltage is breakdown along the fuse surface due to thermal emission before evaporation occurs. All these effects are strongly dependent on the conductor material as well as the surrounding medium. Fig. 2-33 shows the results obtained by U. Schwarz of the time developments of electrically exploded Cu wires with different surrounding media[22]. In the figure the time zero corresponds to a state where the fuse is already molten ($T \approx 1000\mu s$). Constrictions have developed at the wire connections, leading to formation of arcs at $t = 30\mu s + T$. During the following 100 μs, the wire is breaking along the entire length and successive arcs are formed at the breaking positions. After the entire length of the wire is bridged by an arc the voltage drops, which means that the resistance of the fuse is decreasing.

Embedding the wire in a liquid dielectric with low compressibility[23] delays the expansion of evaporated material and suppresses inhomogeneities, which leads to early arcing in the vapor, as a result the onset of peripheral discharges can thus be delayed, or even prevented. The effect of a liquid dielectric on the development of

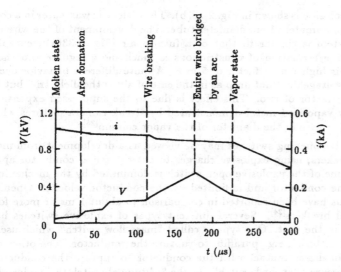

Fig.2-33(a) Current and voltage variations with time for a Cu wire in 1 atm air after 1ms current flow duration (Cu wire: $\phi = 0.25$ mm, $l = 6$ cm) [22].

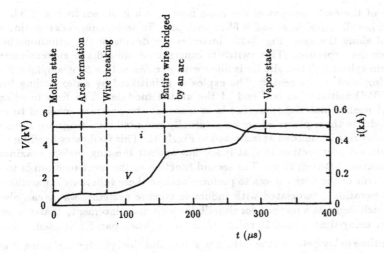

Fig.2-33(b) Current and voltage variations with time for a Cu wire in a water-filled tube after 1ms current flow duration (Cu wire: $\phi = 0.25$ mm, $l = 6$ cm) [22].

the fuse explosion is shown in Fig. 2-33(b). The dielectric was water in a completely closed, approximately 1 cm diameter tube. The development of the wire explosion in this system is similar to that of a fuse in air [Fig. 2-33(a)], except for the fact that the formation of constrictions is much more regular and the voltage generated is higher by a factor of 5 to 6. A quite different behavior occurs after this. The voltage does not drop as in the case of air as the dielectric, but increases by another factor of two. This effect is due to the suppressed expansion of the completely vaporized metal demonstrated by streak camera pictures which show a smaller increase in the diameter of the vapor column[23].

Explosive opening switches may be viewed as a development from mechanical circuit breakers, using explosive charges to cut the metal conductor apart. The opening time of the explosive open switch is dominated by the mechanical movement of the conductor and is limited by the conductor velocity. Opening times about 10 μs have been reported in comparison to about 1 ms or more for typical mechanical breakers[24]. Several different types of explosive switches have been developed in the past. One type is called "mass flow switch" which uses an arc-quenching material, e.g. paraffin, to rupture the conductor. The other type uses explosive in direct contact with the conductor to rupture the conductor and is called "pressure quenched switch". In the 3rd type of switch the explosive and the conductor are mixed together to form a single element of explosive conductor. In addition, there are a few more, but they are mostly still in the early development stage and we shall not elaborate.

One of the early designs of the mass flow switch is shown in Fig. 2-34. The switch is in cylindrical form and is filled with paraffin and having a detonating cord arranged along the axis. Fig. 2-35 illustrates the details of the situations before and after the explosion. This switch is constructed around the current-carrying aluminum cylinder. The cylinder is filled with paraffin with a 50 grain/foot PETN detonating cord in the center. The explosive is initiated by an exploding bridge wire (EBW) initiator at one end of the switch module. The first function of the inert pusher material (paraffin) is to transmit the pressure required to burst the cylinder in those places where no externally mounted steel rings are emplaced. Every other ring is rounded as indicated in Fig. 2-34. This facilitates bending of the aluminum cylinder section so that it lies flat against the ring surface maximizing the separation between rings. The second function of the inert medium is to cool the arc. One of the best choices to perform both of these functions is paraffin. For pulsed operations, associated with inductive storage systems, water can also be used, which suggests a method for simplifying switch replacement. Water provides a shorter delay time, about 50 rather than 70 μs, when paraffin is used.

The time delay between the initiation signal and the cylinder rupturing is given in Table 2-6. This is the time required for the explosive to detonate, compress the paraffin, fill available voids, and rupture the aluminum tube as the appropriate ring dies. The delay time decreases with the weight of the explosive as can be seen from the table. For design purposes, the paraffin velocity may be calculated approximately from the empirical formula (2-43).

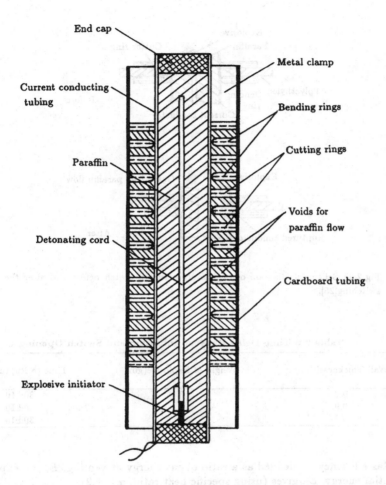

Fig.2-34 Schematic of early design of explosively actuated opening switch assembly[25].

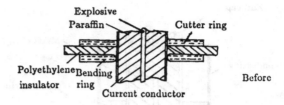

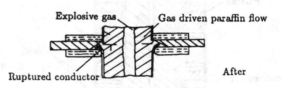

Fig.2-35 Schematic diagram of explosively actuated switch before and after the switching[25].

Table 2-6 Time Delay Between Initiation and Switch Opening

Wall Thickness(mm)	Weight of Explosive(gm)	Time to Rupture(μs)
0.9	3	80±10
0.9	6	50±10
0.9	9	30±10

Gas efficiency, γ, defined as a ratio of gas energy at venting, E_F, to explosive potential energy, E_l, gives (using specific heat ratio, $\gamma = 1.2$);

$$\eta = \frac{E_F}{E_l} = \left(\frac{V_i}{V_f}\right)^{\gamma-1} = 50\% \qquad (2\text{-}42)$$

The efficiency of transferring the internal energy of the gas to the kinetic motion of a pusher is estimated to be about 50% on the basis of results obtained in a test assembly using similar explosive weight, gas volume, and pusher weight. Thus, the overall efficiency, η_{tot} of 25%, suggests that the pusher velocity, v, given by[26]

$$v = \left[\frac{2\eta_{tot}W}{m}\right]^{0.5} \tag{2-43}$$

is 0.12mm/μs for a set of parameters given in Ref.[26]. The estimated velocity was found to agree with the measured values.

Design of an open switch to meet certain specific requirements is a painful task. As pointed out previously, at present there is no well-developed technology that can be reliably employed to design an open switch with certainty. In many instances, one has to rely extensively on empirical data and trial and error methods. In the following we shall outline the basic procedures for the determination of fuse dimensions, even though there are considerable limitations when they are applied.

If the load in an inductive energy storage system is inductive, say an inductor L_l, then the required length h and cross section A of the fuse switch may be estimated by using the following formulas provided the empirical data for that particular material are known[19]:

$$\left(\frac{Ah}{\epsilon_o}\right)\frac{L_1 + L_l}{L_l} = \frac{b}{k_2}$$

$$h = \frac{L_l}{L_1 + L_l}\left(\frac{C_1 L_1}{2}\right)^{1/4}\left(\frac{k_1}{\sqrt{k_2}}\right)^2 V_o b a^{1/2} \tag{2-44}$$

In the above expressions:

ϵ_o is the initial total energy$= (1/2)I_o^2 L_1$;
V_o is the initial applied Voltage at C_1;
L_1 and L_l are storage and load inductors;
k_1 and k_2 are two parameters having the following values:

$$1 < k_1 < 3$$

$$1 < k_2 < 3$$

a and b are material constants, for Ag, Au and Al metals, they are given in Table 2 - 7.

Table 2-7 Values of constants a and b for Ag, Au and Al at 20°C

Metal	a	b
Ag	3.9×10^{16}	4×10^{-11}
Au	3.0×10^{16}	3.5×10^{-11}
Al	2.2×10^{16}	3.5×10^{-11}

The fuse with dimensions A and h so determined from formula (2-44) shall explode under the conditions spelled out by the formula and shall interrupt the circuit properly, (yielding maximum current). As can be seen from Table 2-7, the values of a and b are given for three metals only, if the switch is made from some other metals, then the above method fails and an alternative approach is necessary. For Cu, one may use an empirical formula[27] or graphical method[28] to estimate the required dimensions of the fuse switch provided other circuit parameters are given.

References of Chapter 2

[1] G. J. Rohwein, *IEEE Proc. of the 16th Power Modulator Symposium*, (1984), 200.

[2] R. E. Reinovsky et al., *4th IEEE Pulsed Power Conf.*, (1983), 196.

[3] M. M. Kekez et al., *7th IEEE Pulsed Power Conf.*, (1989), 123.

[4] Y. G. Chen et al., *4th IEEE Pulsed Power Conf.*, (1983), 45.

[5] M. J. Kushner et al., *J. Appl. Phys.* **58**, (1985), 1744.

[6] D. J. Johnson et al., *IEEE Trans.* PS-8, No.3, (1980), 204.

[7] R. Rompe et al., *Zs Physik*, **B 122** H 9-12, (1944).

[8] J. C. Martin, SSWA/JCM/1065/25, 1965; SSWA/ JCM/ 703/ 27, (1970) and SSWA/ JCM/ 704/ 49, (1970).

[9] T. P.Sorensen et al., *J. Appl. Phys.* **48**, (1977), 115.

[10] L. X. Schneider, *4th IEEE Pulsed Power Conf.*, (1983), 202.

[11] Qi Zhang, *Pulsed Power Technology Fundamentals* (in Chinese), (Institute of Electronics, Beijing, 1987), 25.

[12] Operating Manual, *Flash X-ray model 43733A*, (Hewlett-Parkard company, USA).

[13] R. J. Commisso et al., *7th IEEE Pulsed Power Conf.*, (1989), 272.

[14] D. Giorgi et al., *5th IEEE Pulsed Power Conf.*, (1985), 619.

[15] Karl. H. Schoenbach et al., *Proc. of IEEE*, **72** No.8, (1984), 1019.

[16] C. W. Mendel et al., *J. Appl. Phys.* **48**, (1977), 1004.

[17] R. Stringfield et al., *IEEE Trans.* PS-11, (1983), 200.

[18] C. W. Mendel et al., *Rev. Sci. Instrum.* **51**, No.12, (1980), 1641.

[19] C. H. Maisonnier et al., *Rev. Sci. Instrum.* **37**, No.10, (1966), 1380.

[20] A. Guenther, K. Kristiansen and T. Martin, *Opening Switches*, (Plenum Press, New York, 1987).

[21] Y. A. Kotov et al., Pribory i Tekhnika Eksperimenta No.6, (1974), 107.

[22] U. Schwarz, Ph. D. Dissertation. *TU Braunschweig*, (1977), 44.

[23] U. Seydel et al., *Z. Naturforsch* **30a**, (1975), 1166.

[24] R. D. Ford et al., *Rev. Sci. Instrum.* **53**, (1982), 1098.

[25] R. D. Ford and I. M. Vitkovitsky, NRL Memo Report 3561, NRL Washington D. C. (1977).

[26] C. H. Johansson, *Detonics of High Explosives*, (Academic Press, New York, 1970).

[27] E. I. Azarkevich, *Zurnal Tekhniceskoj Fiziki, (in Russian)*, 1, (1973), 141.

[28] L. S. Blokh, *Practical Nomograph, (in Russian)*, (Science Publishing House, Moscow, 1971).

Chapter 3
INSULATION

3-1 Introduction

One of the essential elements in high voltage engineering is electrical insulation. Without proper insulation, no high voltage can be maintained. To achieve proper insulation requires good electrical insulators. What is an electrical insulator? There are many different ways to define it. Fundamentally speaking, an insulator is a material in which the valence electrons of the constituents are relatively tightly attached to their atoms. Under normal electrical stress, these electrons cannot be detached from the atoms and neither can thus move freely. In terms of engineering language, it may be stated as follows: a dielectric insulator is a material that can provide high resistance to the passage of an electrical current. Generally speaking, a good insulator should have a resistivity of no less than 10^{10} ($\Omega \cdot$cm). Table 3-1 shows the resistivities of some common solid and liquid insulators.

Table 3-1 Resistivity ρ at room temperature

Material	$\rho(\Omega \cdot \text{cm})$	Material	$\rho(\Omega \cdot \text{cm})$
Distilled water	$10^5 - 10^6$	Lucite	$10^{15} - 10^{17}$
Castor oil	$10^{12} - 10^{13}$	Quartz	$10^{18} - 10^{19}$
Transformer oil	$10^{12} - 10^{15}$	Teflon	$10^{15} - 10^{18}$
Glass	$10^{14} - 10^{15}$	Polyethylene	$10^{16} - 10^{17}$
Paper	$10^{16} - 10^{17}$	Polymide	$10^{16} - 10^{18}$
Mica	$10^{15} - 10^{16}$	Porylene	$10^{16} - 10^{17}$
Hard rubber	$10^{16} - 10^{17}$		

From Table 3-1, we may see that, the solid insulators in general have relatively higher resistivity than the liquid. However, in most high power pulse applications liquid or gaseous insulators are mostly employed because in solid insulators a volume breakdown results in irreparable damage to the insulating material whereas a breakdown in gas or liquid insulators can be easily cured. Another important feature of gas and liquid insulators is that the breakdown field strength is a time function and follows certain known relationships such as

$$E_B t^{1/3} A^{1/10} = k \tag{3-1}$$

where E_B is the breakdown field strength in MV/cm, t is time in μs, A is the area of the electrode in cm^2 and k is a constant for a given parallel plane electrodes configuration. From Eq.(3-1), we can see that if the system can be charged fast enough, significantly high breakdown field strength may be achieved. Breakdown field strength (sometimes it is also referred to as dielectric strength) is the maximum electric field that an insulator can withstand without breakdown. In high voltage engineering, this is probably one of the most important factors to be concerned with, because with high dielectric strength, the stored electrical energy density can be made significantly high thus permitting the volume of the device to be made considerably small. For example, in an energy storage capacitor, if the dielectric strength of the medium is increased by a factor of two, roughly speaking the volume of the capacitor may be reduced to one quarter of its original size, as the energy density is proportional to the square of the dielectric strength. Although the dielectric strength of a given medium is generally considered to be one of the intrinsic properties of the medium itself, however from Eq.(3-1) we may see that its magnitude is closely influenced by many external factors e.g. the time of charging, the geometrical area etc. Therefore when one quotes the value of the dielectric strength for a given medium, care must be taken to include the specific condition under which it is evaluated, otherwise significant error may be introduced. In the following sections we shall discuss these aspects respectively in a more detailed manner.

Another important measure of the property of an insulating material is its dielectric constant ϵ. From the name we can understand that for a given material it is usually taken as a constant. However the various ways that influence the physical quantities under consideration are rather broad. For example, if high energy density is required, one should use a medium of a large dielectric constant as the energy density is directly proportional to the latter. On the other hand, if high transfer rate of energy is desired, then one should use a medium of a smaller dielectric constant as the E-M wave speed is inversely proportional to the square root of ϵ. Therefore the matter of choice of dielectrics is rather circumstantial and largely dependent on what is the main objective one wants to achieve. The room temperature dielectric constants for some commonly employed insulators in high power pulse applications are given in Table 3-2.

Table 3-2 Room temperature dielectric constants for some common insulators

Material	Dielectric constant	Material	Dielectric constant
Air (1–atm)	1.00053	Glass	4.5–10
N_2 (1–atm)	1.0006	Mica	3–7
SF_6 (1–atm)	1.002	Mylar	2.8
Water	80.4	Teflon	2.8
Castor oil	4.5	Polyvinyl chloride	3.0–3.5
Transformer oil	2.2–2.5	Polyethylene	2.5–2.6
Glycerin	44.0		

In addition to the electrical properties, knowledge about the mechanical and thermal properties of the insulator is also important. Quite often, the insulator has to operate under severe conditions, such as a high power switch operated at repetitive mode, the ability of the insulator to survive the hostile environment is crucial as failure of any one component may jeopardize the whole operation. The following (Table 3-3) are the results of a study to determine the failure modes of several insulators under the condition of high power repetitive discharges. The failure modes include mechanical fracture, softening, deformation, ablation, crazing and current tracking. The materials under test were plastic, quartz, transite and glass.

Table 3-3 Observed failure mode

Plastic	failed from softening and ablation
Transite	failed from current tracking
Quartz	failed from mechanical fracture
Glass	failed from none

3-2 Gaseous Insulation

(a) Dielectric Strength Under Static Voltage.

Electrical breakdown of gaseous insulators under static voltage may be classified into two categories: breakdown in uniform field and in non-uniform field. Gas breakdown in uniform field has been studied quite extensively in the past and its characteristics are relatively well understood. The general behaviors of such breakdown can be reasonably expressed by analytical formulas. For example, the breakdown voltage V_b of an air gap of separation d(cm) at pressure p(bar) in a uniform field may be expressed by

$$V_b(kV) = 6.72(pd)^{1/2} + 24.36(pd) \qquad (3\text{-}2)$$

Except at very low or very high values of the product pd, the above formula has been proven to be reasonably accurate. The calculated values of V_b in the range of 10^{-3} to 10^3 (bar.mm) are shown in Fig 3-1.

Atmospheric air can be used as the basic insulation for various practical applications in high voltage engineering. However, the atmospheric pressure varies considerably in time and locations due to temperature difference, therefore correction of the temperature effect must be taken into consideration when the formula of (3-2) is employed. For practical purposes, it is usually done by using the relative density δ of the gas instead of the gas pressure p, i.e. the expression given in Eq.(3-2) should be written as

$$V_b = 6.72(\delta d)^{1/2} + 24.36(\delta d) \qquad (3\text{-}3)$$

where δ is defined as the ratio of the actual gas density to that at standard conditions ($p = 1.01$ bar, $T = 293$K)

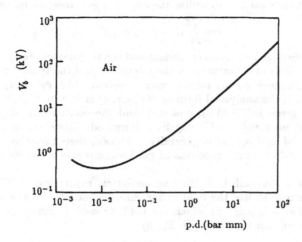

Fig.3-1 Breakdown voltage V_b versus pd values for air in a uniform field.

For gases other than air, the breakdown voltage in a uniform field may be estimated by using formula (3-3) and the relative dielectric strengthes factor γ given in Table 3-4, namely,

$$V_b' \simeq \gamma V_b \qquad (3-4)$$

where V_b' is the breakdown voltage of a gas other than air, and under the same conditions (same p and d) as that of air.

Table 3-4 Relative dielectric strength of several gases

Gas	Relative strength γ
N_2	1.0
CO_2	0.9
SF_6	2.3–2.5
Freon	2.4–2.6
CCl_4	6.3

Description of the breakdown of gaseous insulators in a non-uniform field is more complex than that in the uniform field cases. In this case, there is no general formula of analytical form available for the prediction of the breakdown voltage of a gas insulated spark gap. Instead it has to be worked out on an individual basis for each different electrodes configuration, as the field non-uniformity is very much dependent on the electrode geometry. The only general approach that may be used in determining a relationship between the breakdown voltage and certain parameters of the spark gap (e.g. the pressure p, separation d and the electrodes radius of curvature r etc.) is to utilize the criterion for streamer initiation, i.e.

$$\exp \int_0^d \bar{\alpha} dx = n_e \simeq 10^8 \tag{3-5}$$

where $\bar{\alpha}$ is the effective ionization coefficient and is a function of the applied voltage V, pressure p, radius of curvature r of the electrodes and the spatial coordinate x. n_e is the critical charge density for streamer initiation and is generally assumed to be equal to 10^8. If the analytical form of $\bar{\alpha}(V, p, r, x)$ is known, then by evaluation of the integral given in Eq.(3-5), one may find the relationship $V_b(p, r, d)$ from which the breakdown voltage V_b may be determined. However, in practice, the analytical form of $\bar{\alpha}(V, p, r, x)$ is generally unknown, therefore, the usefulness of the expression (3-5) for the prediction of gas breakdown in a non-uniform field is very limited.

One practical approach in finding an analytical expression of $V_b(p, r, d)$ in a non-uniform field is to use the field enhancement concept. In any non-uniform field there is always a point at which the field strength is maximum. One may express such maximum field strength E_m by

$$E_m = f E_0 \tag{3-6}$$

where E_0 stands for the mean field strength and is defined as

$$E_0 = V/d \tag{3-7}$$

and f is the field enhancement factor and is a function of the spark gap parameters r and d. By taking E_m' as the breakdown field of the gas in question, and combining Eqs. (3-6) and (3-7), one may express the breakdown voltage of the gas in a non-uniform field approximately as

$$V_b = (E_m'/f)d \tag{3-8}$$

where d is the gap separation. Once E_m' and f are both known, the V_b value can be readily calculated from Eq.(3-8). For most practical electrodes configurations, f may be expressed analytically in terms of two parameters r and d, where r is the radius of curvature of the electrodes. Table 3-5, shows f expressions for some common electrodes configurations. For other electrodes configurations, the field enhancement factor f may be obtained by the method suggested in reference [1] or from empirical data.

With the field enhancement factor f is known one still needs to know the value or explicit form of the maximum field E_m' before the breakdown voltage V_b can be determined from Eq. (3-8). There is no simple way to determine the explicit form or the value of E_m'. It can be done only by extensive experimental studies or by traditional field calculations. For the cases of sphere-sphere and parallel cylinders configurations, the E_m' may be approximated by the breakdown fields at the respective surface of the electrodes. When $(d/r) > 1$, for air gaps E_m' may be expressed approximately by[2]

Table 3-5 Field enhancement factor for some
common electrode configurations

Electrode configuration	f factor
Sphere-plane	$0.9(r + d)/r$
Sphere-sphere	$0.9(r + d/2)/r$
Cylinder-plane	$0.9d/r \ln[(r + d)/r]$
Parallel-cylinders	$0.9d/\{2r \ln[(r + d/2)/r]\}$
Perpendicular-cylinders	$d/[2r \ln((r + d/2)/r)]$
Coaxial-cylinders	$d/[r \ln\{(r + d)/r\}]$
Concentric spheres	$(r + d)/r$

$$\text{spheres} \qquad E'_m = 24.6p + 6.7(p/r_s)^{1/2} \qquad\qquad (3\text{-}9)$$

with $r_s = 0.115r$ and

$$\text{cylinder} \qquad E'_m = 24.6p + 6.7(p/r_c)^{1/2} \qquad\qquad (3\text{-}10)$$

with $r_c = 0.23r$

By substituting the E'_m expressions from Eqs.(3-9) and (3-10), and the corresponding f expressions from Table 3-5 into Eq. (3-8), we have the approximate breakdown voltage for sphere-sphere and parallel cylinder air gaps as follows:

$$\text{spheres} \qquad V_b = [d(24.6p + 19.8(p/r)^{1/2})]/[0.9(1 + d/2r)] \qquad (3\text{-}11)$$

$$\text{cylinder} \qquad V_b = [54.6p + 31.1(p/r)^{1/2}]r\ln[((r + d/2)/r] \qquad (3\text{-}12)$$

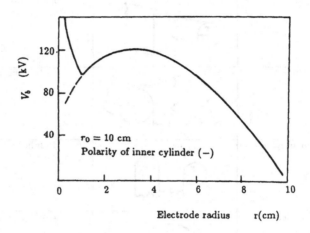

Fig.3-2 The relationship of the breakdown voltage V_b with the inner electrode radius r of the coaxial cylinders in air [6].

In Eqs.(3-11) and (3-12), both d and r are in cm, p in atm and V_b in KV. Breakdown voltages calculated from these equations, may have a few percent of errors depending largely on the d/r ratio. For d/r values less than one, the error may be considerably high. For air gaps of other configurations, analytical formulas like that given in Eqs.(3-11) and (3-12) are not generally available. In these cases one would have to employ other methods, e.g., field calculation, to find the value of the maximum field E'_m and then use Eq. (3-8) to estimate the breakdown voltage

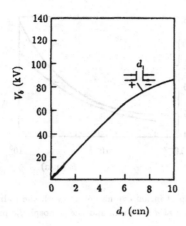

Fig.3-3 The relationship of static breakdown voltage with the separation for point-point air gap[6].

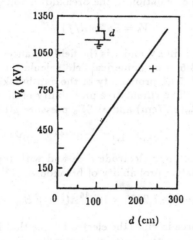

Fig.3-4 The relationship of static breakdown with the separation for rod-plane air gap[6].

V_b, or employ the available empirical data for evaluation. Fig. 3-2 to Fig. 3-5 show the empirical data of air and SF$_6$ breakdown in various electrodes configurations.

If gases other than air are employed, for rough estimations, the corresponding breakdown voltage can be obtained by multiplying the relative strength factor γ

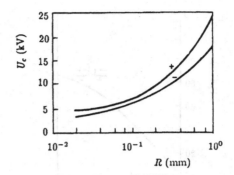

Fig.3-5 The relationship of initial corona voltage with the radius of curvature for point-plane configuration at $d = 20$ mm and one atmospheric pressure of SF_6 [6].

given in Table 3-4 with the corresponding air result. However, for more accurate evaluations, one should always follow the appropriate approach for that particular gas under consideration. For instance, if SF_6 gas is employed, one should use the following formula for the calculation of the breakdown voltage:

$$V_b = 88.7p(d/f) \tag{3-13}$$

p is in atm, d is in cm, V_b is in kV and f is the field enhancement factor which may be obtained from Table 3-5 or by numerical field calculation techniques. Equation (3-13) is the voltage with 50% probability of the gap breaking down and the gap is assumed to be perfect. The cumulative probability of breakdown of such a gap with a maximum field E_m(kV/cm) and at SF_6 pressure p(bar) is[3]

$$Q_1 = 1 - \exp[-1.77 \times 10^{-44}(E_m/p)^{22.2}] \tag{3-14}$$

For imperfect SF_6 gaps of large electrode area and with rough electrode surface, the corresponding cumulative probability of breakdown is[4]

$$Q_2 = 1 - \exp[-7.2435 \times 10^{-16} A(\text{cm}^2)(E_m - E_o)^{6.3}] \tag{3-15}$$

where A is the effective area in cm^2 (the electrode area that is stressed within 90% of the maximum field E_m on the negative electrode), E_o is a parameter may be found from Ref.[4].

   Gas breakdown in a non-uniform field not only depends on pressure p and separation d, but also depends on the voltage polarity. Fig.3-6 shows the positive and negative point-plane gap corona inception characteristics measured in air as a function of gas pressure. Fig.3-7 shows the corona inception characteristics of an identical point plane gap in SF_6 gas. From these results, one can see that in the presure range of 0.1 to 0.5 MPa, the corona inception characteristics under both positive and negative polarities are similar except the inception voltage for the positive polarity is considerably higher than that for the negative polarity. This common feature holds for both air and SF_6 gaps. Under similar condition, the corona breakdown characteristics, however, are distinctively different. For example, under positive polarity in air the breakdown voltage increases with pressure of up to about 7 bar, then a sudden drop follows[5]. Under negative polarity there is no such sudden drop in the breakdown strength in this pressure range. Breakdown characteristics of SF_6 display similar behaviors as that for air. In this case the corona breakdown region under positive voltage is limited to pressures below approximately 1.5 bar. Fig. 3-5 shows the polarity effect with respect to the radius of curvature of the point electrode in a point-plane gap for SF_6. It can be seen from the figure, under positive point voltage, the corona inception voltage is also higher than that for negative polarity and the characteristic is similar to that shown in Fig.3-7.

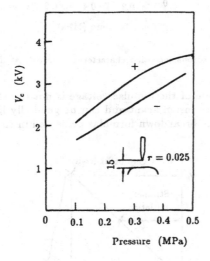

Fig.3-6 Point-plane corona inception characteristics in air at $d = 15$ mm and $r = 0.025$ mm) [6].

(b) Gas Breakdown Under Pulse Voltage

The breakdown criterion for gases under pulse voltages is more complex than that for static breakdown. For pulse breakdown, there are two effects need to be taken into consideration. One is the so called statistical time delay required for the initial electrons to be produced. The other effect is the time required for the subsequent process of streamer formation and propagation. If the field is

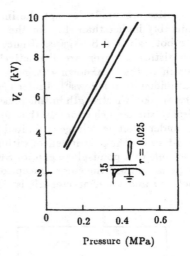

Fig.3-7 Point plane corona inception characteristics in SF_6 at $d = 15$ mm and $r = 0.025$ mm [6].

uniform, and the risetime of the impulse voltage is not too short (>few hundred ns), these two effects are insignificant and can be generally ignored. Under such circumstances, the static breakdown formulas presented in the above section may

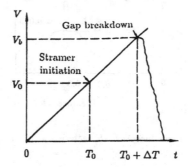

Fig.3-8 Modification of streamer initiation.

still be applicable to estimate the approximate breakdown voltage. However, if the field is not uniform or the impulse is very short (few tens of ns or less), then the appropriate pulse approach will be necessary. In this case, specification of both the risetime and peak voltage of the impulse is necessary, because for a given gas spark gap, it may breakdown at many different voltages if the risetimes of the impulses are different.

At present, there is no general formula which can completely describe the breakdown behaviors of all gas spark gaps under pulsed charging conditions. For large gaps with divergent fields, an approximate relation may be employed to estimate the breakdown voltage of the gas gap[7], i.e.

$$V_b = K_\pm \times p^n d^{5/6} t^{-1/6} \tag{3-16}$$

where V_b is the breakdown voltage in kV, p is in atmosphere, d is the gap separation in cm and t is the time in μs at which the voltage exceeds 89% of the breakdown voltage, K and n are constants and depending on gases as well as on polarities. Their values are given in Table 3-6.

Table 3-6 K and n value

	Air	Freon	SF$_6$
K_+	22	36	44
K_-	22	60	72
n	0.6	0.4	0.4

The above formula applies for pressures from one to five atmospheres or so. For air the time dependency disappears for times longer than the order of 1 μs for negative voltage pulses, and several hundred μs for positive pulses. For gases other than those given in Table 3-6, one may still use the relationship of (3-16) and combining the function with γ values given in Table 3-4 to estimate the respective breakdown voltage.

In establishing the relationship given in Eq.(3-16), it was assumed that the streamer started immediately after the field was applied, i.e. $t = 0$ and then the streamer propagated across the gap. The time t in Eq.(3-16) is therefore the time required for the streamer to propagate across the gap. Such an assumption is rather crude as initiation of the streamer usually requires some time after the field is applied. Therefore certain modification of the relationship given in Eq.(3-16) is needed. One such modification[3] assuming the initiation of streamer takes place when the voltage reaches the d.c. breakdown voltage (Fig. 3-8). In the figure, the gap voltage is assumed to increase linearly with time and from $t = 0$, $V = 0$ to $t = T_o$ and $V = V_o$ where the d.c. breakdown occurs (the streamer is launched). With such modification, the time ΔT may be approximated as

$$\Delta T = [(114 d^{5/6} p^{0.56})/V_b]^6 \tag{3-17}$$

where ΔT is in μs and represents the streamer transient time after the streamer is initiated at the d.c. breakdown voltage. Once the impulse voltage ramp is known, from Eqs.(3-7) and (3-17), one may by means of a computer determine the breakdown voltage V_b. Fig. 3-9 and Fig. 3-10 show respectively the relationships between the breakdown voltages and the gap separations in air and SF$_6$ under specific impulse voltages of microseconds duration.

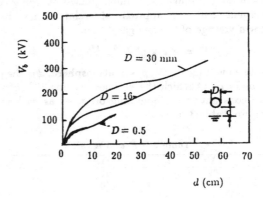

Fig.3-9 The relationship between the breakdown voltage and the gap separation at the wire-plane air gap under the (+) lightning impulse voltage [6].

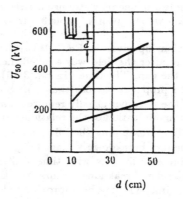

Fig.3-10 The relationship between 50% breakdown voltage and the separation, under very non-uniform field, at 1 atm SF_6 and with impulse voltage of (1.5/40) μs [6]

3-3 Liquid Insulation

(a) General Remark

The general state of knowledge on the electrical breakdown in liquids is less advanced than that for the cases of gas breakdown, in particular, information concerning d.c. breakdown in liquid is even less. Although many aspects of liquid breakdown have been studied in the past, the findings and conclusions obtained from these studies cannot be reconciled and in many instances they are contradictory. Based on the published data, some review articles on the subject have been reported periodically and the more recent ones may be found in the references [8–10]. But these reports are incomprehensive and can only furnish some very limited information concerning d.c. breakdown in liquid dielectrics.

Study of pulsed breakdown in liquids, on the other hand, has received considerably more attention in recent years, due to the great demands from the pulsed power field for the developments of liquid dielectric devices, e.g. high power water transmission lines. The main motivation for the development of liquid dielectric devices for the pulsed power applications is their mechanical or structural compatibility with the high power pulse system which is usually required to store high density of electrical energy and to withstand high voltages. In principle, there is a large variety of liquids that can be used as insulators in the high voltage environment, for practical usage however only a small number of liquids is suitable. The most widely employed liquids in high power pulse applications are transformer oil and water. Because oil represents a type of liquid with large dielectric strength and small dielectric constant, making it suitable for low inductance devices. Water is attractive due to different reasons. Mainly because it has a very high dielectric constant ($\epsilon = 80$) making it very useful for high density energy storage. Besides, a third liquid of water/ethylene glycol mixture (about 5 % water) has been studied[11]. Results indicate that such a liquid has the properties somewhere between oil and water. Its dielectric constant is close to that of water and its resistivity is close to that of oil, therefore it has an intrinsic time constant ($\tau = K_o - \epsilon\rho$) in the range of millisecond. Moreover, its breakdown strength in the millisecond range is much higher than that for pure water. All these properties suggest that it has great potential as an insulator in pulsed power applications.

(b) Breakdown Strength

Breakdown characteristics of oil and water under pulsed voltages have been studied quite extensively under various conditions such as the dielectric strength, time duration and effect of impurities etc. For uniform fields, the breakdown voltage of a given spark gap of electrode area A (cm^2) may be approximately expressed in MV by[7]

$$V_b = K_\pm A^{-1/10} dt^{-1/3} \tag{3-18}$$

where $K_\pm$ is a constant and depends on liquids and polarity (Table 3-7), d in cm is the gap separation and t in μs is the effective time and is defined as the duration for which the voltage reaches 63% of its maximum value.

**Table 3-7 Breakdown constant for coaxial
pulse lines and center conductors**

	Oil	Water
K_+	0.5	0.3
K_-	0.5	0.6

The relationship given in Eq.(3-18) is valid for both oil and water however water has to be constantly deionized so that its resistivity can be maintained at a level above 1 megohm centimeter. Effects of rep-rate pulse voltage on the breakdown of mineral oil has been reported to be insignificant up to 100 pps. However, when the oil was subjected to a 5 μs pulse at 1000 pps, its breakdown strength was found to be about 20% lower than that obtained at 10 and 100 pps[12].

For non-uniform fields the breakdown field for water may be approximately expressed by[13]

$$F_{\pm} = \alpha K_{\pm} A^{n\pm} t^{-1/3} \tag{3-19}$$

in Eq.(3-19), $F_{\pm}$ is the breakdown field in MV/cm and is polarity dependent, α is a function of the field enhancement factor f and is defined as

$$\alpha = 1 + 0.12(f-1)^{1/2} \tag{3-20}$$

$K_{\pm}$ and $n_{\pm}$ are constants as shown in Table 3-8.

Table 3-8 $K_{\pm}$ and $n_{\pm}$ constants

K_+	K_-	n_+	n_-
0.23	0.557	-0.058	-0.069

A is the electrode area within 90% of the peak electrical field and is in cm^2, t is the effective time as defined previously in Eq. (3-18).

The breakdown strength of water may be enhanced significantly by increasing the pressure[12]. However, analytical expression of pressure dependence for water, similar to that given in Eqs. (3-2), (3-11) and (3-12) for gases has not been established. The application of such a method to enhance the breakdown strength, at present still has to rely on empirical data. As for other liquids, it is also not clear whether the relationship given in Eq.(3-18) holds or not. Therefore, in these cases one may have to rely on empirical data as well. Table 3-9 shows the breakdown strength of several liquids other than water and transformer oil.

Table 3-9 Breakdown strength of some highly purified liquids

Liquid	Strength (MV/cm)
Hexane	1.1–1.3
Benzene	1.1
Silicone	1.0–1.2
Glycerine*	0.1
Nitrogen	1.6–1.9
CuSO₄, solution*	3.0–3.6
Castor oil	3.9

* under μs pulse condition, others under d.c. conditions.

3-4 Solid Insulation

(a) Breakdown Characteristics

Solid insulation forms an integral part of most high voltage facilities. The solid materials not only provide the mechanical support for the conducting parts but also insulate the conductors from one another. Therefore a knowledge concerning the electrical breakdown mechanism of solid insulators is of great importance. However, the breakdown mechanism of solids is not so well understood as that of the gases, although many investigations have been carried out in the past. One of the complications is that there are several distinct mechanisms that may contribute to a breakdown and the breakdown strength of a solid insulator changes with the time of voltage application. The basic breakdown mechanisms of solids so far have been identified and studied include the following:

(1) Intrinsic Breakdown — it is generally considered to be due to electrons in the insulator gain sufficient energy from the applied field to cross the forbidden energy gap from the valence to the conduction band. Intrinsic breakdown is usually accomplished in the order of 10^{-8} s.

(2) Streamer Breakdown —the mechanism is conceptually similar to the streamer mechanism in gases, namely when the avalanche exceeds certain critical size breakdown will occur. Transit time is usually short.

(3) Thermal breakdown —when an insulator is stressed electrically, because of currents and dielectric losses due to polarization, heat is continuously generated within the insulator. If the rate of heat generation exceeds the rate of heat losses, then the insulator will undergo thermal breakdown. Such breakdown takes place usually in a much slower time scale.

(4) Erosion Breakdown—Insulators often contain voids or cavities within the material. These cavities are usually filled with gas or liquid medium of lower breakdown strength than the solid insulator. Accordingly, under normal working stress of the insulator the voltage across the cavities may exceed the breakdown value hence initiating breakdown in the cavities.

(5) Breakdown due to tracking and other causes—Tracking is the formation of some permanent conducting paths, usually carbon, along the insulator surface due to degradation of the insulator or other causes. Besides tracking, there are also electromechanical breakdown and treeing breakdown etc.

(b)Breakdown Strength

The breakdown of a solid insulator may be contributed from one or a combination of several different mechanisms mentioned above depending on the time duration of application of the applied field. Fig. 3-11 shows the general behaviors of breakdown strength in solids with time of stressing. For example, in the time range of 10^{-6} to 10^{-9} sec, breakdown of a typical solid insulator may be summarized as follows: at a very early time, intrinsic breakdown field is high. As the time of the applied field extends, streamers leading to lowered breakdown begin to develop. At an even later time, thermal erosion and tracking take place, usually in association with repetitive charging of the insulator. Besides, the breakdown strength is also volume dependent. Fig. 3-12 shows the dependence of breakdown strength on the dielectric volume for several common solid insulators. The streamers formed in solids during breakdown usually have a short transit time, the breakdown field is nearly independent of the pulse duration (down to a few nanoseconds). A practical expression for the breakdown field is given by[14].

$$E = KM^{-1/10} \qquad (3\text{-}21)$$

where the field E is in MV/cm, the volume M is in cm^3, K is a constant and its values is given in Table 3-10 .

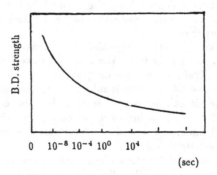

Fig.3-11 General behavior of breakdown strength in solids with time of stressing.

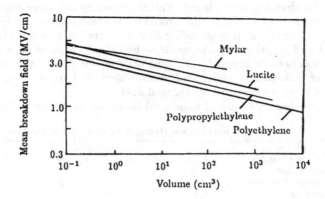

Fig.3-12 Dependence of mean breakdown field of four dielectric sheets on dielectric volume subjected to electric field stress [14].

Table 3-10 Values of K

Material	K
polyethylene	2.5
teflon	2.5
polypropylethelene	2.9
perspex	3.3
mylar(thick)	3.6

For thin sheets, the standard deviation of the breakdown field decreases and the breakdown strength becomes nearly independent of the volume. For repeated pulse, there is a reduction in breakdown field for solid insulators. The number of charging cycles before failure occurs may be expressed by

$$\text{Life} = (E_b/E_p)^8 \tag{3-22}$$

where E_b is the breakdown field and E_p is the operating field at which the life is wanted. For example, a piece of polythene of volume 10^4 cm^3 has a mean pulse breakdown field of 1 MV/cm for a single pulse. However, if it is required to operate for 1000 pulses, then the operating field should be reduced to about 0.5MV/cm.

For dc charged solids, several effects combine to alter the breakdown strength and usually but not invariably, lower it. In some plastic conduction currents can heat the plastic and cause run away thermal degradation. Chemical corrosion from surface tracking can cause degradation of the breakdown fields, as can mechanical flow under electrostatic forces. All of these effects vanish in pulse charged systems. For instance in polythene, dc charging can yield fields some 20 to 30% lower than that when pulse changing is employed. This is caused by enhanced conduction in the regions containing the defect that originates the breakdown. However, if the voltage is rapidly reversed, this annealing charge separation now adds to the field on the defect and polythene can be made to break with a pulse reversal of only 30% when it has been dc charged. The time scale of this charge and annealing is of the order of millisecond. Mylar and perspex show little or none of this effect and hence are to be preferred for dc charged devices.

The dc breakdown strength of some solid insulators are given in Table 3-11.

Table 3-11 Breakdown strength of some solid insulators

material	strength(KV/cm)
Alumina Ceramics	79–118
Boron Nitride	354–551
Electrical Ceramics	22–118
Borosilicate glass	1772
Steatite	57–110
Nylon	118–157
Depolymerized Rubber	142–150

* Samples are all 3.18 mm thick.

Fig. 3-13 shows the short time breakdown strength of polythene versus temperature. Fig. 3-14 is the breakdown behaviors of mylar with respect to the thickness. Fig. 3-15 is the result of polypropylene breakdown voltage as a function of voltage risetime.

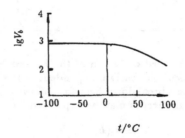

Fig.3-13 Short time breakdown strength of 1 mm polythene versus temperature [6].

3-5 Surface Flashover

(a) Surface Flashover Characteristics

Surface Flashover is commonly referred to as the phenomenon of electrical discharge along or near an insulator surface. There are many different forms of surface flashover depending on the geometrical configurations of the system as

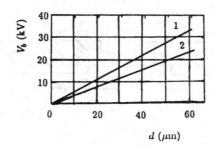

Fig.3-14 The breakdown behaviors of mylar with respect to the thickness at the ambient temperature [6].

    1–D.C. voltage, uniform field;

    2–A.C. voltage, uniform field.

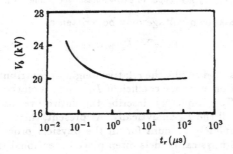

Fig.3-15 Breakdown voltage V_b versus voltage risetime t_r for a piece of 60 μm thick polypropylene.

well as the physical condition under which the phenomenon takes place. Generally speaking, they may be summarized into four basic types of surface flashover as shown in Fig. 3-16.

The configuration in Fig.3-16(a) is the most simple one in which the insulator surface is parallel to the electric field. The general effect of the insulator in the gap is to lower the breakdown voltage. Fig. 3-17 is a comparison of the breakdown voltages between the cases having insulator (curves 2 and 3) and cases without it (curve 1). The physical mechanisms that are responsible for the effect of lowering breakdown voltage are rather complex. It is generally assumed that, not a single, but a number of factors or processes are collectively responsible for the effect and each of these processes has a functional dependence on some of the system

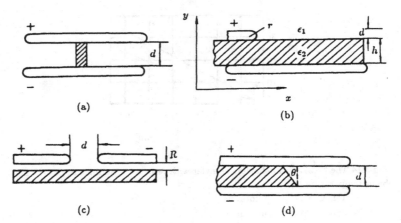

Fig.3-16 Typical types of surface flashover configuration.

parameters. The breakdown voltage may be written as

$$V_b = \sum_n F_n(\text{process})_n \qquad (3\text{-}23)$$

where the coefficients F_n describe the relative weight of contribution of each process for a given configuration and each coefficient F_n depends on the system parameters in a different way. To completely describe the entire breakdown processes, one would have to know all the F_n analytically. At present, there is no comprehensive theory that can completely account for all the physical processes involved. The existing theories and physical models often place exceptional significance only on one or two processes and set the other F_n's to be nearly zero such that the variation of these particular coefficients F_n's with the system parameters can be studied. For example, surface flashover is known to depend on many processes such as the influences due to surface property, surface roughness, surface cleanness and system configuration etc., however, many actual studies e.g. Fig. 3-17 only place emphasis on one process, namely the influence due to surface property and ignore the rest. Such types of investigations have obvious limitations, but at the present time , this is probably the only practical way to tackle a complex problem such as surface flashover. In the following we shall discuss some of the limited models concerning surface flashover. The ultimate goal, of course, is to arrive at a physical model that can describe the contribution of every conceivable process to the surface flashover event.

(b)Basic Processes and Physical Models

(1) The configuration (a) shown in Fig.3-16 is the simplest and most widely studied one among the four. Its breakdown behavior represents the typical surface

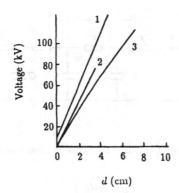

Fig.3-17 The relationship between breakdown voltage and d shown in Fig.3-16(a).
(1) breakdown voltage in the air; (2) flashover voltage with a paraffin cylinder; (3)
flashover voltage with a ceramic cylinder [6].

flashover phenomenon, and it is the most common type of flashover encountered in
many practical instances. However, the physical mechanism that is responsible for
the breakdown behaviors is still not quite clear. For example, it is well known that
the presence of the insulator in the gap reduces the breakdown voltage, but the
questions concerning how the breakdown voltage is reduced and by what physical
mechanism are still poorly understood. All we know is that the presence of the
insulator in the gap modifies the electrical field, but only qualitatively. Quantitatively, at present there is virtually no theory or model that can completely
and precisely describe even one of the very basic phenomena—reduction of the
breakdown voltage. Therefore, until further progress is made in this respect, we
may have to rely on a qualitative approach to treat the problem, even though it
appears to be rather crude.

(2) In the surface flashover configuration shown in Fig. 3-16(b), it is assumed
that no breakdown occurs through the insulator and only a long discharge along
the surface takes place until the energy source is exhausted or until the streamer
bridges the two electrodes. An earlier study on the subject was made on the basis
of equivalent electric circuit[15]. It was assumed that the discharge configuration
may be represented by an equivalent circuit (Fig. 3-18).

The relationship between the current i and the voltage drop U along the surface
is assumed to be

$$\left.\begin{array}{l} \delta i/\delta x = \sigma U + C\delta U/\delta t \\ \delta U/\delta x = i\rho \end{array}\right\} \tag{3-24}$$

where σ is the conductivity per unit area, C is the capacitance per unit area and
ρ is surface resistivity. From these Eqs., the breakdown voltage V_b was deduced

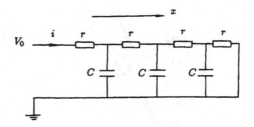

Fig.3-18 Equivalent circuit of configuration shown in Fig.3-16(b).

to be

$$V_b \approx k/(C)^{1/2} \qquad (3\text{-}25)$$

where k is a constant determined by the configuration, applied voltage frequency and the surface property. From Fig. 3-18, we can see that the degree of approximation made in this approach was severe, therefore, the usefulness of Eq.(3-25) is rather limited.

A more recent study on the same subject was made on the basis of electrical field enhancement mechanism[16]. In this approach, the components electric field E_x and E_y were analyzed as a function of the ratios d/h, ϵ_2/ϵ_1 and r/h. Here ϵ_2 and ϵ_1 refers to the respective dielectric constants of the insulator surface and the surrounding environment. The results indicated that, for $d/h = 0.5 \sim 1$, the field character at distances from the surface greater than the electrode radius r is $E_x > E_y$. In this case, the E_x component determines the process of initiating the discharge. Along the surface, E_x reaches a maximum where the breakdown was initiated. For $d/h = 0.3$, the electric field along the surface from the triple point is $E_y >> E_x$, essentially independent of the geometry. Resulting from accumulation of surface charges of the same sign as that of the high voltage electrode, the E_y component may be significantly weakened. However, in the region $d/h > 0.2$ to 0.7, both E_x and E_y may be enhanced by increasing the ratio of ϵ_2/ϵ_1. In summary, this model suggests that the breakdown behaviors are closely related to the ratios of d/h and ϵ_2/ϵ_1. However, these relationships can only be seen graphically but not analytically. Moreover, the model did not explain how the insulator surface influences the breakdown voltage in a precise way.

(3) The breakdown behaviors with the configuration shown in Fig. 3-16 (c) was studied specifically to determine the effect of the insulator surface on the breakdown voltage and channel formation[17]. The basic assumption in the model which they developed is that when the product of electron density and electron drift velocity reaches a certain critical value, breakdown will occur. This is a sharp contrast in comparing it with the conventional models in which the electron density alone was always taken to be the breakdown critical parameter. In this model, the breakdown voltage was seen to be determined by two competing processes: production of electron from ionization and loss of electrons due to diffusion, which

of the two processes is predominant depending on many factors. The analytical expression for predicating the breakdown voltage V_b is as follows

$$V_b = a\{(G/R^2)\mathrm{J}_0(2.405r/R)\exp(t - t_0/\tau) + n_0\}^{-1} \qquad (3\text{-}26)$$

where a and G are constants for a given system, J_0 is the zero order Bessel function, r is a spatial coordinate, R is the distance between the electrodes and the insulator surface, t_0 is the time at which the electron density at the surface reaches n_0, t is the time at which breakdown occurs. τ is the time constant and is defined by

$$1/\tau = f_e - (2.405/R)^2 D \qquad (3\text{-}27)$$

where f_e is the net rate of electron production and D is the electron diffusion coefficient.

From Eq.(3-26), we can see that V_b is R dependent and there exists a minimum breakdown voltage as R increases from zero. This clearly demonstrates the effect of the insulator surface (through R) on the breakdown voltage. The significance is that it shows the effect of the insulator is not only to modify the field based on geometrical considerations, but also to modify the electron transport rate and the volumetric electron density. In fact, this is virtually the only model that shows a direct interaction between the insulator surface and the discharge environment. Fig. 3-19 is the result, obtained by using a flat quartz insulator and the experimental set-up shown in Fig. 3-16(c), The circles are the observed results and the curve is the calculated one from Eq. (3-26).

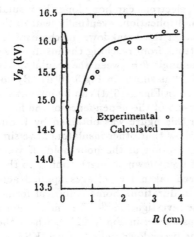

Fig.3-19 Experimental and calculated breakdown voltage as a function of R for a flat quartz substrate [17].

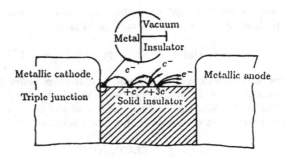

Fig.3-20 A schematic illustration of the physical regime associated with a 'solid' high voltage insulator: also shown is the 'hopping' mode by which electrons travel from cathode to anode.

(c) Surface Flashover in Vacuum

The configuration given in Fig.3-16(d) is frequently used in a vacuum environment. It is therefore more appropriate to start with the subject of surface flashover in a vacuum environment for the discussion. Qualitatively, the basic process in surface flashover in a vacuum environment may be illustrated by using Fig.3-20. Initially, some primary electrons are produced due to field emission from the high field region, e.g. the triple junction region. Subsequently, secondary electrons are produced as a result of bombardment of the insulator surface by the primary electrons initiated from the triple junction region. Under the influence of the applied electric field, electrons can hop along the insulator surface towards the anode and after many multiplications eventually lead to the final breakdown. For more detailed discussion on the breakdown mechanism, see reference [18]. One way to prevent the electrons from reaching the insulator surface hence from multiplying, is to have some angle θ between the insulator surface and the direction of the applied electric field as shown in Fig.3-16(d). By so doing, the breakdown voltage of the set-up given in Fig. 3-16(d) can be increased considerably. Fig. 3-21 represents the typical result of the dependence of the flashover field on the angle θ. The result shown in this figure was obtained by I. Smith et al. from epoxy cone[23]. Investigations show that other insulators have similar dependence on the angle θ[19]. With the exception of the poor hold-off voltage of glass, the other materials have values of breakdown strength similar to that of the epoxy resin.

These data have been taken for voltages applied across insulators in short pulses, 30 ns and 1 μs in duration. Some empirical formulas for practical usage were deduced from these investigation[20]. For one type of applications e.g. in the graded insulator stack as shown in Fig. 3-22(a), where the angle is $\theta = 45°$, the empirical formula for the breakdown field E_b (in MKS units).

$$E_b = K_{\pm}t^{-1/6}A^{-1/10} \tag{3-28}$$

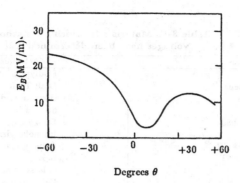

Fig.3-21 Surface flashover voltage dependence on the angle θ [23].

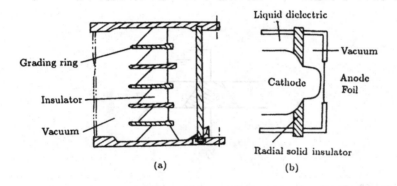

Fig.3-22 Vacuum insulators: (a) graded vacuum insulator stack and (b) radial solid insulator.

where K is a constant depending on the material and the angle θ, K_+ is for positive polarity and K_- is for negative polarity, t is the time during which the insulator effective area A is stressed to more than 89% of the maximum field strength. The K values for lucite are $K_+ = 7 \times 10^5$, and $K_- = 4.2 \times 10^5$[21]. Under the condition shown in Fig.3-23, surface flashover voltages of various dielectric materials have been determined by O. Milton[22]. The materials studied are listed in Table 3-12. The approximate values of $K_\pm$ for these materials may be obtained by using the data presented in Ref.[22].

Table 3-12 Materials to which the flashover
voltages have been determined [22]

Materials	Materials
Lucite	Diallyphallate
Polyethylene	Diall TS-5
7740 glass	Boron nitride
Teflon	65-35 PZT
Teflon etched	Epoxy glass
828-DFA	Phenolic glass
828-B, MACH	Polystyrene
828-B, CAST	Lexan
SRIR	C lecstyrene
828-DFA, MICA	Lead borate
Nylon phenolic	Coated alumina
Metal ionomer	Cast nylon
Reynoids Al_2O_3	MC901
Coors Al_2O_3	Polyurethene foam
SE9090 6E	Silicone rubber

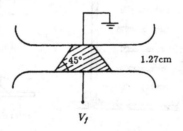

Fig.3-23 Experimental set-up employed in determining flashover voltage for the materials listed in Table 3-12, with $V_f = 5$ μs pulse of $\pm$ polarities [22].

For other types of application, e.g. for low inductance radial insulator as shown in Fig. 3-22(b), the empirical formulas are[23]

$$\left.\begin{array}{l} E_s \leq 5 \times 10^5 t^{-1/6} A^{-1/8.5} \\ E_T \leq 1.6 \times 10^6 t^{-1/6} A^{-1/8} \\ E_p < 5 \times 10^6 \end{array}\right\} \qquad (3\text{-}29)$$

where E_s and E_T are the greatest electric field parallel to the insulator surface and greatest value of the total field, and E_p is the field parallel to the surface near the cathode or anode intersections with the insulator, respectively.

The relationship given in Eq.(3-28) is for unipolar pulses, if the applied voltage is a bipolar pulse, then the corresponding formula after modification is[21]

$$E = K'_{\pm} t^{-1/2} A^{-1/10}, \qquad \text{for} \quad E > E_{\min} \tag{3-30}$$

where E in kV/cm, is the peak electric field. The insulator was subjected to E before flashover occurs, t, in μs, is the time measured from the half amplitude point on the first prepulse (Fig. 3-24) to the point that flashover occurs, $K'_{\pm}$ are constants, $E_{\min}$ is the minimum flashover field for long duration waveforms. The value of K'_{+} for these materials are: lucite $K'_{+} = 33$, polystyrene $K'_{+} = 42$, cast epoxy $K'_{+} = 37$. The $E_{\min}$ with 300 ns pulse for polystyrene of 20.8 mm thick is $E_{\min} = 33$ kV/cm, $E_{\min}$ for lucite and cast epoxy are not known. The K'_{-} values have not been determined.

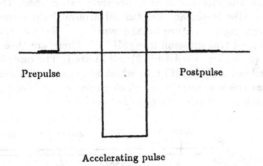

Fig.3-24 Bipolar waveform used in flashover experiment.

3-6 Other Insulating Techniques for Pulsed Power Applications

(a) Insulation by Magnetic Method

In the configuration shown in Fig. 3-16 (d), when a magnetic field B is present such that $E \times B$ is pointing away from the insulator surface (Fig. 3-25), then the electron avalanche may be pulled away from the surface hence the electron multiplication along the surface as indicated in Fig. 3-20 may be reduced[24,25]. If the magnetic field, either self-produced or externally applied, is sufficiently large, the surface flashover strength may be significantly enhanced. Such a technique is usually referred to as magnetic flashover inhibition (MFI). The basic requirement for the MFI process to be effective is (in MKS unit)

$$E \leq 2.1 \times 10^7 B \tag{3-31}$$

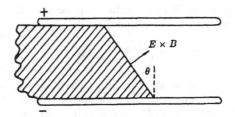

Fig.3-25 Magnetic Flashover inhibition.

where E is the average electric field between the conductors and B is either the self or applied magnetic field.

Similarly, in a vacuum transmission line (Fig. 3-26) if a self or applied magnetic field is present and the electron flow is properly established, then the electrons emitted from the cathode surface are turned around before they can reach the anode, hence preventing premature breakdown[26-28]. Such a structure is called a magnetically insulated transmission line (MITL). There are three types of MITL, (1) constant flux, (2) load limited and (3) self-limited. The one shown in Fig. 3-26 is the second type load limited MITL in which the current flowing to the anode in region D_1 provides the magnetic field which prevents the electrons emitted from the cathode to reach the anode in region D_2.

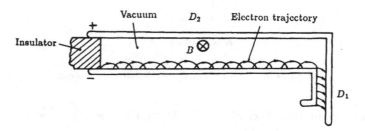

Fig.3-26 Load limited MITL [26].

The essential condition for establishing the proper electron flow is

$$I_e \geq I_a[1 + (eV_a)/(mc^2)]^{-1} \qquad (3\text{-}32)$$

where I_a is the anode current and $I_a = I_c + I_e$, I_c is the cathode current, I_e is the electron flow in the vacuum, V_a is the anode voltage, e and m are the electron charge and mass, respectively and c is the speed of light. For practical applications, the essential criterion may be roughly expressed as[29]

$$V_m = d^2 t^{-2} K^2 \qquad (3\text{-}33)$$

where V_m is the maximum voltage the MITL can withstand, d is the separation between the two conductors, t is the time at which breakdown occurs and K is constant but dependent on the voltage shape. For $V = $ const. $K = (2^{1/2})/c$, for $V = $ triangular pulse with risetime and fall time both equal to $(1/2)\tau$, $K = 2/c$, for a voltage of exponential form $V = V_o \exp(\alpha t)$, $V = V_{\max}$, at $t = \tau$, then $K = (\alpha\tau)/c$, here c is another constant and $c = 190 \pm 65$.

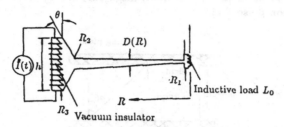

Fig.3-27 In a typical application of short pulse, pulsed power, the generator produces a dI/dl in the vacuum inductor. The power flows through the vacuum insulator and the magnetically self-insulated transmission line to an inductive load [29].

In a high power field, quite often, power in the TW (10^{12} w) range has to be transported to the load through a long vacuum transmission line, such as in the case of inertial confinement fusion facilities. The combined vacuum interface flashover and the vacuum gap closure constitute the power flow problem and frequently limit the power that is available at the load. By employing the MFI and MITL techniques, such problems can be lessened significantly. Fig. 3-27 is one of the typical applications of such techniques reported by J. P. Vanderender[29]. In the figure, all the quantities are assumed to be known except h and $D(R)$. Here $D(R)$ is the MITL gap profile and is given by

$$D(R) = (\tau/K)[V_m(R)]^{1/2} \tag{3-34}$$

where τ is the time at which breakdown occurs. K is a constant and defined previously in Eq.(3-33), and $V_m(R)$ is the maximum voltage at position R on the MITL, and

$$V_m(R) = \{10^{-7}(\pi/K)\dot{I}_m \ln(R/R_1) + (L_o\dot{I}_m)^{1/2}\}^2 \tag{3-35}$$

where $\dot{I}_m = (dI/dt)_{\max}$ is assumed to be given. As the structure shown in Fig. 3-27 employs both the MFI and MITL techniques, it has to meet both of the requirements. These requirements are:

$$E = 2.1 \times 10^7 B \tag{3-36}$$

with

$$\left.\begin{array}{l} E = V_3/h, \quad B = \mu_o I/(2\pi R_2) \\ V_3 = \dot{I}[L_3 + V_m(R_2)/\dot{I}_m] \\ L_3 = 5 \times 10^{-8}(h^2/R_3)\tan\theta \end{array}\right\} \tag{3-37}$$

$$(V_3/h)\tau^{1/6}(2\pi R_2 h)^{1/10} \leq 7 \times 10^5 \qquad (3\text{-}38)$$

From these equations, one may determine the values of the parameters h, T, V_3, and L_3. Such calculations for a triangular $I(t)$ waveform with the total risetime and fall time equal to 10^{-7} s have been carried out by the same author. The calculations indicate that multi-megavolt power sources will be required to drive inductive loads in vacuum at multi-megajoule energy levels in 200 ns with the most common pulse shape.

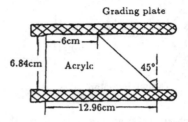

Fig.3-28 Typical dimensions of an insulator stack component employed in various insulator stacks.

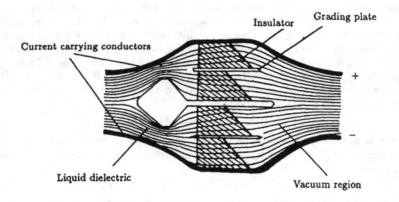

Fig.3-29 Equipotential distribution in regions separated by a vacuum insulator stack.

Fig.3-28 shows the basic structure of a component of the vacuum insulator stack shown in Fig. 3-27. Fig.3-29 shows the equipotential distribution in regions separated by a vacuum insulator stack.

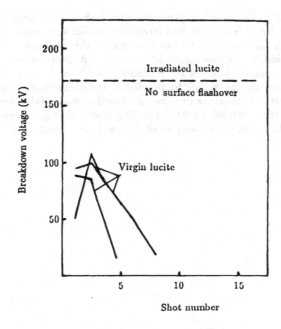

Fig.3-30 Data plots for virgin and treated lucite [31].

(b) Enhancement of Flashover Strength by Chemical Treatment

Vacuum flashover strength of certain insulators may be increased significantly by proper treatment of the insulator surface with discharge irradiation[30]. In a recent report[31], lucite surface was treated in vacuum with irradiations, and after the treatment, it was found by the authors that the vacuum flashover strength of the insulator increased by about 70%. Fig.3-30 shows the results obtained by Hatfield et al.[31] for both the treated and fresh (untreated) lucite samples. Analysis of the treated surface indicates that the surface structure of the insulator has been modified and a new layer of carbon-oxygen compound was formed on top of the original surface. It is believed that such physical modification of the surface is responsible for the enhancement of flashover strength in vacuum. However, the physical mechanism that is responsible for the modification of the surface is still not known. An attempt was made to artificially produce such modified layer under controlled conditions by vacuum deposition method. But the efforts did not produce the expected results and in fact the surface flashover strength of the produced surface was found to be even lower than the untreated ones. The exact reasons are not known. Another mystery of the treatment effect is that not all the treated insulators show similar behaviours even when they were treated under identical conditions. For example, the vacuum flashover strength of Nylon was found to decrease by 10% after it was irradiated under the same conditions. So, at present, it may be too early to say how useful this technique is in terms of practical application to the enhancement of vacuum flashover strength. However, judging from the results obtained by Hatfield et al.[31] and reproduced in Fig.3-30, it certainly has the potential to provide a complementary method for the improvement of surface flashover in vacuum. Another potentially useful approach is to encapsulate the electrodes with insulating material, e.g., acrylic. By means of this method, the surface flashover strength has been enhanced significantly[32].

References of Chapter 3

[1] J. W. Boag, *PIEE*, **100-3** Pt.IV (1953) 63.

[2] J. C. Martin, *Dielectric Strength Note 11*, SSWA/ JCM/ 6611/ (1966) 106.

[3] T. H. Martin, *5th IEEE Pulsed Power Conf.*, (1985) 74.

[4] T.Nitta et al, *IEEE Trans. on PAS*, **PAS-93**, (1974) 623.

[5] E. Kuffel and W.S. Zaengl, *High Voltage Engineering Fundamentals*, (Pergamon Press, Oxford, 1984).

[6] Zhu Deheng, Yan Zhang, *High Voltage Electrical Insulation*, (Tsinghua University Press, Beijing, 1992).

[7] J. C. Martin, *Nanosecond Pulse Technique*, (SSWA/JCM/704/49, 1970) 14.

[8] Z. Krasucki, *Breakdown of Commercial Liquid and Liquid Solid Dielectrics*, High Voltage Technology (Alston), (Oxford Univ. Press, 1968) p.129.

[9] A. A. Zaky and R. Hawley, *Conduction and Breakdown in Mineral Oils*, (Pergamon Press, Oxford, 1973).

[10] T. J. Gallagher, *Simple Delectric Liquids*, (Clarendon Press, Oxford, 1975).

[11] D. B. Fenneman, *J. Appl. Phys*, **53**, (1982) 8961.

[12] R. Kraus, et al., *IEEE 7th Pulsed Power Conf.*, (1989) 332.

[13] W. B. Moore, et al., *IEEE 5th Pulsed Power Conf.*, (1985) 315.

[14] I. D. Smith and J. C. Martin, Note SSWA/IDS/6610/105 (1965).

[15] W. Claussnitzer, *Arch. f. Electr.*, **49** (1965) 271.

[16] P. N. Dashuk et al., *Sov. Phys. Tech. Phys.*, **24**, (1979) 687.

[17] S. T. Pai et al., *J. Appl. Phys.*, **53** (1982) 8583.

[18] A. S. Pillai et al., *J. Appl. Phys*, **53** (1982) 2983.

[19] R. Hawley, *Vacuum*, **18** (1968) 383.

[20] J. C. Martin, *Fast Pulse Vacuum Flashover*, SSWA/JCM/713/157 (Aldermaston, U.K. 1971).

[21] W. K. Tucker et al., *5th IEEE Pulsed Power Conf.*, (1985) 323.

[22] O. Milton, *IEEE Trans. Electrical Insulation*, **EI-7**, (1972) 9.

[23] I. Smith et al., *15th IEEE Power Modulator Symp.*, (1982), 160.

[24] K. D. Bergeron, *J. Appl. Phys*, **48** (1977) 3073.

[25] J. P. Vandevender et al, *8th Int. Symp. on Discharge and Electrical Insulation in Vacuum*, (1978) E-1.

[26] J. M. Credon, *J. Appl. Phys.*, **48** (1977) 1070.

[27] J. P. Vandevender, *J. Appl. Phys.*, **50** (1979) 3928.

[28] C. W. Mendel et al., *Laser and Particle Beams*, **1**, **Part 3** (1983) 311.

[29] J. P. Vanderender, *3rd IEEE Pulsed Power Conf.*, (1981) 248.

[30] G. L. Jackson et al., *IEEE Trans on Electrical Insulation*, **EI-18** (1982) 310.

[31] L. L. Hatfield et al., *5th IEEE Pulsed Power Conf.*, (1985) 311.

[32] G. A. Tripoli et al., *7th IEEE Pulsed Power Conf.*, (1989) 820.

CHAPTER 4
TRANSMISSION LINE

4-1 Fundamentals of Transmission Line

(a) Basic Concept

What is a transmission line? Briefly speaking, any two conductors between which voltage is applied and which can transmit electricity may be considered as a transmission line. In many instances, there is no clear-cut distinction between a transmission line and an ordinary electric circuit. In those cases one can treat the two conductors either as a transmission line or as an ordinary electric circuit. However, in many other cases, there is a significant difference between the two. Then one must treat them properly, otherwise serious errors may be introduced into the result when it is analyzed. Whether the two conductors should be treated as a transmission line or as an electric circuit depends mainly on two factors, the length of the conductors and the characteristics of the applied voltage. More specifically, it depends on the ratio of the length of the conductors and the wavelength of the applied voltage. If the wavelength is very long compared to the length of the conductors, it can be treated as an electric circuit. Otherwise it should be considered as a transmission line. Let's use the following example to illustrate why this is so.

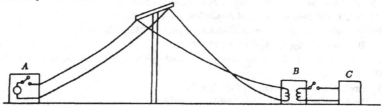

Fig.4-1 Simplified power transmission system consisting of both transmission line and electric circuit.

In Fig.4-1, let the blocks A, B and C represent respectively a transmitter, a voltage amplifier and a receiver. The distance between points A and B is 60 km and that between B and C is 0.6m. It is further assumed that when the switch at location A is closed, a voltage signal in the form of $V = V_m \sin 2\pi f t$ is generated at A. Here f is the frequency of the signal and is assumed to be 1 kHz which is typical for audio frequency signals. The corresponding wavelength is 300 km and is comparable to the distance between A and B. This voltage signal, however, will not appear instantaneously at point B, as a certain time is required for the signal to travel from points A to B. Even the signal travels with the speed of light (3×10^8 m/s), it would take 0.2 ms to reach there. During this period, the voltage at point

A changes its magnitude continuously. Assume that at $t = 0$, the voltage at point A is zero as shown in Fig.4-2, 0.2 ms later this voltage of zero magnitude will reach point B. Meanwhile the voltage at point A has increased to 0.95 V_m. That is when the voltage at the transmitter is already $0.95V_m$, the voltage appearing at the amplifier remains zero. Anywhere along the line between points A and B, the voltage V will be $0.95V_m > V > 0$. Thus there is a voltage (potential) difference along the line as a result of finite time required for the voltage produced at point A to reach point B. Because of this voltage difference along the line, it is necessary to consider the conductors connecting point A and B as a transmission line. Besides it is also essential in order to maintain proper phase relations when more than one generators are connected into the transmission system. Now let us assume that, at $t = t_0$, the switch at B is closed, because the distance between B and C is so small, the voltage at point B will appear almost instantaneously at the receiver (except a voltage drop across any series resistors or impedance that may be present along the line). For all practical purposes one can also ignore the phase difference if more than one source is present. In this case, the conductors connecting points B and the receiver can be treated as an electric circuit. However, if the frequency of the voltage is increased to 10^8 Hz (typical FM radio frequency) the corresponding wavelength is 3 m which is comparable to the conductor length between points B and C, then it must be treated as a transmission line because the voltage difference along the line between points B and C will exist. This is why the antenna and lead in wires connecting to a radio are usually treated as transmission lines, otherwise distortion of the signal will occur. From this example, one can see that as pointed out previously, when the wavelength of the applied voltage is comparable to or shorter than the length of the conductors, the conductors must be treated as a transmission line.

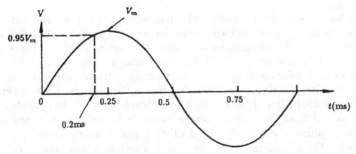

Fig.4-2 Voltage appears at point A as a function of time t.

In the high power pulse applications, the applied voltage is usually in the form of a single or multiple pulses. It is well known that a single pulse is composed of an infinite number of Fourier components with the amplitudes of the spectrum decreasing for higher frequencies. To deal with such a voltage pulse consisting of an infinite number of different frequencies is obviously very difficult if not altogether impossible. An alternative is to employ the empirical formula to estimate the upper-frequency f of the pulse, i.e.

$$f = \frac{K}{t_r} \qquad\qquad (4\text{-}1)$$

where t_r is the risetime of the pulse and k is a constant between 0.35 and 0.45 depending on the shape of the pulse[1]. Since the risetimes of the pulses involved in this field mostly are in the range of subnanosecond to tens of microsecond, from Eq.(4-1) one can see that the corresponding upper frequency would be in the range of a fraction to several hundred mega cycle/sec. The length of the conductors employed in high power pulse system is usually between a fraction of a meter to tens of meters. So these conductors must be treated as transmission lines, because the equivalent wavelength of the pulse in most cases is comparable to the length of the conductors.

In a high power pulse system, the transmission line is one of the basic elements and can serve several important functions. It can produce very fast risetime pulses with the pulse width determined by the length of the line. It can transmit a large amount of pulsed power i.e. large current and voltage simultaneously with high fidelity and fixed propagation time to a target tens of meters away. With proper switching, transmission lines can be used to re-shape the form of the incoming pulse with the pulse width determined by the length of the transmission line. When they are used for this purpose, they are referred to as pulse forming lines (PFL). Transmission line can also be used as intermediate store (IS). As the name implies, it is used to temporarily store the incoming energy from the main energy store such as a Marx generator into a transmission line. When they are used for such purposes, they are usually placed between the main energy store and the pulse forming line.

The above description classifies the transmission lines on the basis of their functions. On the basis of the basic structure, they can also be classified as simple transmission lines, double transmission lines (Blumlein lines), multiple (stacked) transmission lines and magnetically insulated transmission lines (MITL) though this type of transmission lines is operated on an entirely different principle. As the name implies, the simple transmission line is the simplest both in terms of its structure and operation. It is simple to construct and easy to operate, however, the drawback is that the output voltage can only be half of the charging one. If high voltage pulse is required, use of other types of transmission lines is more appropriate. The Blumlein line is a typical double line which may be conceptually considered as two simple transmission lines connected in such a way that they are charged in parallel configuration and discharged in series. With proper termina-tion, in principle it can produce output voltage as high as the charging one. It is also easy to operate and relatively more reliable. For these reasons, Blumlein line is considered to be one of the most popular devices in high power pulse ap-plications. When several transmission lines are stacked together, it is called the stacked transmission line. With proper switching and termination, this type of device can produce considerably higher voltage or current than other types can. If high voltage pulse is desired, the individual lines are stacked together in series discharge configuration, if high current pulse is required then they are stacked together in parallel discharge configuration. The magnetically insulated transmission line is structured on an entirely different principle. It utilizes the magnetic field to inhibit the electrons from reaching the other conductor of the transmission line so that good insulation between the two conductors may be achieved. For this reason it can be constructed in a very compact form, yet it is still capable of transmitting a very high power to a tiny target a far distance away. In addition to the above classification, transmission lines can also be classified in terms of their geometric configurations. In this way they may be classified as coaxial (cylindrical) line, parallel cylindrical line, strip line, helical line and tapered line. A detailed discussion of each type of transmission lines mentioned above shall be given in later sections. Before that, let's first discuss some fundamental principles and theoretical formulations concerning transmission lines in general.

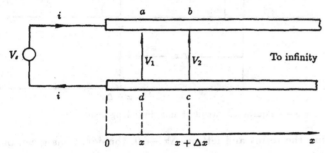

Fig.4-3 An infinite long parallel conductor transmission line.

(b) Fundamental Principles and Theoretical Formulations

In order to facilitate the formulation, it is advantageous to use the equivalent circuit of the transmission line. Since any passive circuit can be composed only of combinations of resistive, capacitive and inductive elements, the equivalent circuit of a transmission line must also contain only combinations of these. Let us consider two very long parallel conductors which are suspended in air as shown in Fig.4-3. Assume that the wavelength of the applied voltage is shorter than the total length of the line but much longer than the segment ab shown in the figure, so that the section ab can be considered as an electric circuit. There are various possible ways to draw an equivalent circuit for the transmission line shown in Fig.4-3. One of the commonly employed equivalent circuits is shown in Fig.4-4 in which R denotes the series resistance, L the series inductance, R' the shunt resistance, C the shunt capacitance and all of them in general are functions of x and expressed in per unit length values, i.e. R in ohms per unit length, L in henries per unit length etc. If the length of the line is l, the total series resistance in general is $\int_0^l R dx$ and for uniform line R =constant, it is reduced to lR. Applying Kirchhoff's law around the loop abcd shown in Fig.4-4 and letting $1/R = G$, the conductance per unit length, we can obtain

$$V_1 = R\Delta x i_1 + L\Delta x \frac{\partial i_1}{\partial t} + V_2 \qquad (4\text{-}2)$$

and

$$i_1 = i_2 + G\Delta x V_2 + C\Delta x \frac{\partial V_2}{\partial t} \qquad (4\text{-}3)$$

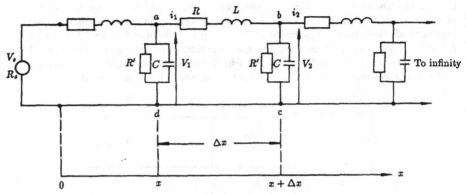

Fig.4-4 Equivalent circuit of an infinite long transmission line.

By rearranging the terms and letting $\Delta x \to 0$, the equations given in (4-2) and (4-3) become[2]

$$\frac{\partial V}{\partial x} = -Ri - L\frac{\partial i}{\partial t} \qquad (4\text{-}4)$$

$$\frac{\partial i}{\partial x} = -GV - C\frac{\partial V}{\partial t} \qquad (4\text{-}5)$$

where R, L, G and C are constants for uniform transmission line only. The general form of Eqs.(4-4) and (4-5) in principle can be solved to obtain $V(x,t)$ and $i(x,t)$ in analytic forms. If so, one can carry out various analyses for the transmission line precisely. In practice, however, the functional forms of the parameters R, L, G and C are seldom given. Even if they are given, it does not necessarily imply that Eqs.(4-4) and (4-5) are readily solvable. So, in most practical cases, the parameters R, L, G and C are treated as constants, then from Eqs.(4-4) and (4-5) one can obtain

$$\frac{\partial^2 V}{\partial x^2} = RGV + (RC + LG)\frac{\partial V}{\partial t} + LC\frac{\partial^2 V}{\partial t^2} \qquad (4\text{-}6)$$

$$\frac{\partial^2 i}{\partial x^2} = RGi + (RC + LG)\frac{\partial i}{\partial t} + LC\frac{\partial^2 i}{\partial t^2} \qquad (4\text{-}7)$$

which are the fundamental governing equations for a uniform transmission line. The solutions of equations (4-6) and (4-7) may have many different forms. The physically meaningful solutions for this case can be written respectively as

$$V = Ae^{j(\omega t - kx)} + Be^{j(\omega t + kx)} \tag{4-8}$$

$$i = A'e^{j(\omega t - kx)} + B'e^{j(\omega t + kx)} \tag{4-9}$$

where $j = \sqrt{-1}$, A, B, A' and B' are constants to be determined by the boundary conditions. In the expressions, ω is the frequency of the waves, k is the propagation constant and is related to other parameters by the relation[3]

$$k^2 = \omega^2 LC - RG - j\omega(RC + LG) \tag{4-10}$$

Both of the solutions given in equations (4-8) and (4-9) represent superpositions of two travelling waves, one going to the $+x$ direction and the other going to the $-x$ direction. That is in a uniform transmission line, both the voltage and current in general are composed of two travelling waves going respectively in opposite directions. The wave going in the $+x$ direction is initiated at the input terminal due to the source or reflection, and that going in the $-x$ direction is initiated at the load (or any discontinuity along the line) due to reflection.

In order to determine the constants A, B, A' and B' of Eqs.(4-8) and (4-9), it is essential to specify the boundary conditions. Let us denote the characteristic impedance of the transmission line by Z_0, the load impedance by Z_l, the length of the transmission line by l, and assuming that the internal impedance of the source $R_s = Z_0$, then from Eqs.(4-8) and (4-9) and using the boundary condition at $t = 0$, $x = 0$, $V = V_0$ and $i = i_0$ we have

$$\frac{V_0}{i_0} = \frac{A + B}{A' + B'} = Z_0 \tag{4-11}$$

and

$$A + B = V_s - i_0 R_s \tag{4-12}$$

Further from the boundary condition at $t = t$, $x = l$, $V = V_l = i_l Z_l$ we get

$$Z_l = \frac{Ae^{-jkl} + Be^{jkl}}{A'e^{-jkl} + B'e^{jkl}} \tag{4-13}$$

From (4-11) and (4-12), by using $R_s = Z_0$, $i_0 = A' + B'$, we obtain

$$A + B = V_s/2 \tag{4-14}$$

and

$$Z_0(A' + B') = V_s/2 \tag{4-15}$$

therefore

$$A + B = A'Z_0 + B'Z_0 \tag{4-16}$$

From Eq.(4-16) we have

$$A = A'Z_0 \quad \text{and} \quad B = B'Z_0 \tag{4-17}$$

where A and A' are the amplitudes of waves travelling in the $+x$ direction, while B and B' are those travelling in the $-x$ direction. But A and B represent the respective incident and reflected voltage waves, while A' and B' represent the respective incident and reflected current. If the polarities of the currents such as A' and B' are chosen as positive when they are going in the $+x$ direction in the top conductor, then the relations given in Eq.(4-17) should be written as

$$A = A'Z_0, \quad B = -B'Z_0 \tag{4-18}$$

because a positive B implies it is going in the $-x$ direction where as a positive B' means it is going to the $+x$ direction in the top conductor. Put relations (4-18) into Eq.(4-12) we obtain

$$A = V_s/2 \tag{4-19}$$

Put Eqs.(4-18), (4-19) in Eq.(4-13), we have

$$B = \frac{1}{2}V_s e^{-j2kl}\frac{(Z_l - Z_0)}{(Z_l + Z_0)} \tag{4-20}$$

where $Z_0 = [(R + j\omega L)/(G + j\omega C)]^{1/2}$, is the input impedance of an infinite long or matched line which may be obtained from Eqs.(4-4), (4-10), (4-2) and by setting $\partial/\partial t = j\omega$ in Eq.(4-4). By substituting A, B, A' and B' respectively from Eqs.(4-19), (4-20) and (4-18) into Eqs.(4-8) and (4-9), we finally obtain the general expressions in analytical forms of the voltage and current in a uniform transmission line as follows

$$V = \frac{1}{2}V_s e^{j\omega t}\left\{e^{-jkz} + \frac{(Z_l - Z_0)}{(Z_l + Z_0)}e^{j(kz-2kl)}\right\} \tag{4-21}$$

$$i = \frac{1}{2}\frac{V_s}{Z_0}e^{j\omega t}\left\{e^{-jkz} - \frac{(Z_l - Z_0)}{(Z_l + Z_0)}e^{j(kz-2kl)}\right\} \tag{4-22}$$

where k is given by Eq.(4-10). From the above expressions one can see that the voltage V and current i in general are functions of time t and spatial co-ordinate x and their behaviors are determined by various factors which include the parameters $V_s, \omega, Z_l, l, R, L, G$ and C etc. If the values of these parameters are known, one can readily use Eqs.(4-21) and (4-22) to carry out various analyses and evaluations for the properties of the transmission line. When applying these equations, one should remember that the condition $R_s = Z_0$ must be satisfied, otherwise modifications of the expressions given in Eqs.(4-21) and (4-22) are required. As an example, let us use Eqs.(4-21) and (4-22) to carry out an analysis for an open transmission line ($Z_l = \infty$) subjected to a step rise voltage i.e., $t < 0$, $V = 0$ and $t \geq 0$, $V = V_s$.

Here we have $\omega = 0$, $k = -j\sqrt{RG}$, $Z_l = \infty$. Assuming that all other parameters are known, from Eqs.(4-21) and (4-22), we obtain

$$V = \frac{1}{2}V_s \left\{ e^{-\sqrt{GR}x} + e^{\sqrt{GR}(x-2l)} \right\} \tag{4-23}$$

$$i = \frac{1}{2}\frac{V_s}{Z_0} \left\{ e^{-\sqrt{GR}x} - e^{\sqrt{GR}(x-2l)} \right\} \tag{4-24}$$

When setting $x = 0$, we get the voltage and current at the input end

$$V_0 = \frac{1}{2}V_s \left(1 + e^{-2\sqrt{GR}l}\right) \tag{4-25}$$

$$i_0 = \frac{1}{2}\frac{V_s}{Z_0} \left(1 - e^{-2\sqrt{GR}l}\right) \tag{4-26}$$

The second term in the brackets represents the reflected component. For lossless lines, i.e., $R = 0$, $G = 0$, $V_0 = V_s$ and i_0 become zero. For an infinite long line i.e. $l = \infty$, then $V_0 = V_s/2$ and $i_0 = V_s/(2Z_0)$. At the output end i.e. the load end, i_l always is zero, and the voltage $V_l = V_s \exp -\sqrt{GR}l$. For a lossless line $V_l = V_s$ and for an infinite long line with loss $V_l = 0$.

When a transmission line is terminated with an impedance different from the characteristic impedance of the transmission line, (improper termination) reflection of the voltage and current pulse at the termination will take place. The voltage reflection coefficient ρ_v is defined as the ratio of the negatively going voltage wave to the positively going voltage wave at $x = l$. From Eq.(4-21), by dividing the second term with the first, one can obtain

$$\rho_v = \frac{Z_l - Z_0}{Z_l + Z_0} \tag{4-27}$$

The voltage transmission coefficient is defined as the ratio of the total voltage to the positively going voltage at $x = l$, which gives $T_v = 1 + \rho_v$. The current reflection coefficient is defined as the ratio of the negatively going to the positively going current wave at $x = l$. From Eq.(4-22), by means of the same procedure, one can obtain

$$\rho_i = \frac{Z_0 - Z_l}{Z_0 + Z_l} = -\rho_v \tag{4-28}$$

Thus, the current reflection coefficient is just the negative of the voltage reflection coefficient. It can be readily seen that

$$0 \leq |\rho_v| = |\rho_i| \leq 1 \tag{4-29}$$

Similarly we can obtain the current transmission coefficient as $T_i = 1 + \rho_i$. It is also clear from Eqs.(4-27) and (4-28) that when $Z_l = Z_0$, there will be no reflection to occur as both ρ_v and ρ_i vanish in this case.

In a previous section (Eq.(4-20)), we have mentioned that the characteristic impedance of a transmission line may be expressed as

$$Z_0 = \sqrt{\frac{R + j\omega L}{G + j\omega C}} \tag{4-30}$$

If $R = G = 0$, then $Z_0 = \sqrt{L/C}$. These expressions in fact are valid only for special cases, namely an infinite long line or a properly terminated line i.e., $Z_l = Z_0$. The characteristic impedance of a uniform transmission line in general is a function of x which can be obtained by dividing Eq.(4-21) with Eq.(4-22), namely

$$Z_x = \frac{V_x}{i_x} = Z_0 \frac{(1 + \rho_v e^{-2jkl} e^{2jkx})}{(1 - \rho_v e^{-2jkl} e^{2jkx})} \tag{4-31}$$

For an improperly terminated ($Z_l \neq Z_0$) transmission line of length l, the input impedance Z_{in} can be readily obtained from the above expression when x is set to zero.

$$Z_{\text{in}} = Z_0 \left(\frac{1 + \rho_v e^{-2jkl}}{1 - \rho_v e^{-2jkl}} \right) \tag{4-32}$$

If the line is lossless (G=R=0), then $k = \omega\sqrt{LC}$, Eq.(4-32) becomes

$$Z_{\text{in}} = Z_0 \left(\frac{1 + \rho_v e^{-2j\omega\sqrt{LC}l}}{1 - \rho_v e^{-2j\omega\sqrt{LC}l}} \right) \tag{4-33}$$

If the line is shorted so that $Z_l = 0$, then the input impedance reduces to

$$Z_{\text{in}} = Z_0 \tanh(j\omega\sqrt{LC}l) \tag{4-34}$$

Since the propagation velocity of the wave is $1/\sqrt{LC}$, one can set $\omega\sqrt{LC} = 2\pi/\lambda$, then Eq.(4-34) becomes

$$Z_{\text{in}} = jZ_0 \tan \frac{2\pi}{\lambda} l \tag{4-35}$$

If $l = \lambda/4$, then $Z_{\text{in}} = jZ_0 \tan\left(\frac{\pi}{2}\right) = \infty$ that means the impedance of a $l = \lambda/4$, lossless line, and short circuited at its load end, looks like an infinite impedance to the input source. Such a property has important applications. Similarly, if the output terminal of a quarter-wavelength line is kept open circuit i.e. $Z_l = \infty$, its input impedance becomes $Z_{\text{in}} = -jZ_0 \cot\left(\frac{\pi}{2}\right) = 0$. Then the line appears to be a short circuit to the source generator which also has numerous applications.

(c) Basic Consideration for Pulsed Voltage Input

It is known from wave theory that the velocity of a sinusoidal wave is equal to the ratio of the frequency ω to the propagation constant k, i.e. $v = \omega/k$. If k has a linear dispersion relation with ω, e.g. $k = \omega\sqrt{LC}$, v will be independent of the frequency. On the other hand, in the situation where a non-linear dispersion relation holds, such as that given in Eq.(4-10), waves with different frequencies will travel at different speeds. In the preceding discussions, the applied voltage (the source) has been treated as a sinusoidal wave of single frequency. Therefore the incident and reflected waves both travel with the same speed along the transmission line. In the cases of pulsed voltage, however, both the incident and reflected waves are composed of an infinite number of different frequencies and each frequency component travels along the line with a different speed. To carry out an analysis for such a situation in the general sense, e.g., to analyze the phase and amplitude relations between the various frequency components, under the condition given in Eq.(4-10), is obviously quite cumbersome and difficult. Some approximation therefore is necessary. One practical approach is to treat the transmission line as an ideal line, i.e. uniform, lossless and dispersionless. Under such conditions, all waves will travel along the line with identical speed regardless of what their frequencies are. In other words the various frequency components of a pulsed voltage will travel along the line with identical speed hence they will all arrive at the load end at the same time. When they are added together, it will give the same waveform as that at the input end. For this reason, the formulas derived in the preceding sections, such as Eqs.(4-21) and (4-22), can still be used to ideal with transmission lines under pulse voltage, except in this case, the expressions given in Eqs.(4-21) and (4-22) represent one of the corresponding frequency components in the Fourier frequency spectrum and the relation $k = \omega \sqrt{LC}$ should be employed.

A complete Fourier frequency spectrum includes the component of zero frequency ($\omega = 0$). This component also travels along the line (if it is an ideal line) with the same speed as that of the rest of the components. Therefore, if one uses this component to carry out analysis for a transmission line under pulse voltage, it will not lose the generality and special features of the latter. When $\omega = 0$, we known from the relation $k = \omega \sqrt{LC}$, that k also vanishes. Substituting these values into Eqs.(4-8) and (4-9) and replacing the constants A, B, A' and B' respectively by a set of new notations V_i, V_r, I_i and I_r, we get

$$V = V_i + V_r \qquad (4\text{-}36)$$

$$i = I_i + I_r \qquad (4\text{-}37)$$

The above discussion provides two important information: (1) the zero-frequency component is composed of two constant terms therefore it can be analyzed much more easily. (2) the behavior of the zero-frequency component is similar to that of a step-function input of zero rise-time, and it can be treated as a step-function voltage of magnitude $(V_i + V_r)$ or a step-function of current $(I_i + I_r)$. This information suggests a convenient way to study the behaviors of a transmission line under pulse voltage. In the following sections, we shall present our discussions on this basis.

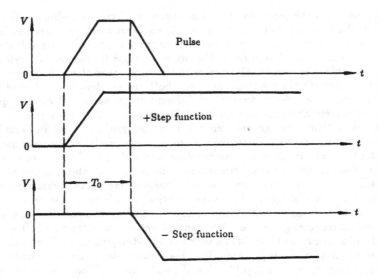

Fig.4-5 Single pulse obtained from superposition of two step functions.

4-2 Ideal Line Response Under Step-Function or Pulse Input

(a) Step Function Input With Zero Rise-time

Since pulse can be constructed from the superposition of two step functions, one positive and one negative and are delayed in time with respect to each other by an amount of T_0 as shown in Fig.4-5, study of the step function response of an ideal line is of great importance. By proper superposition of the reflected waveforms of the step function one can readily obtain, from knowledge of the step function, the behavior of the pulse input. If a transmission line is properly terminated at both ends i.e. $Z_s = Z_0$, and $Z_l = Z_0$, no reflection at either end will take place. Then the input waveform, whether it is a step function of zero or non-zero risetime, a narrow or wide pulse, will retain its original form along the line. When the line is improperly terminated at both ends, i.e. $Z_s \neq Z_0$, and $Z_l \neq Z_0$ multiple reflections between the two ends of the line will occur. Under such situations, the reflected waveforms can be very different for different input waveforms. For example, in an improperly terminated transmission line, an input of step function of zero risetime may result in a quite different waveform in comparison with that for an input of step function with non-zero risetime. For these reasons, one should be careful when treating transmission lines with improper terminations or discontinuities along the line. In the following we shall first discuss how to deal with the situations involving input of step function having non-zero risetime, then proceed to the cases where zero rise-time are involved and finally to discuss the various aspects concerning inputs of single pulse.

The basic formulas employed for the study of the behavior of step function input to an ideal transmission line are rather simple. They are from Eqs. (4-36) and (4-37)

$$V = V_i + V_r \tag{4-38}$$

$$i = I_i + I_r \tag{4-39}$$

where V_i and I_i are the voltage and current initiated at the input end, V_r and I_r are the corresponding ones initiated at the load end due to reflection, respectively. They are related by the relationships:

$$I_i = V_i/Z_0, \qquad I_r = -(V_r/Z_0) \tag{4-40}$$

Their polarities and direction of travel along the line are shown in Fig.4-6. The time for the wave to travel from the input end ($x = 0$) to the load end ($x = l$) in an ideal line of length l is:

$$T_0 = l\sqrt{LC} \tag{4-41}$$

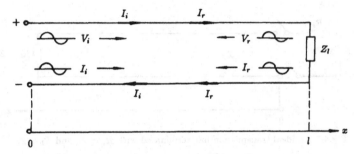

Fig.4-6 Polarities and directions of travel of voltage and current waves.

From Eqs.(4-21) and (4-22), one can further express Eqs.(4-38) and (4-39) in more convenient forms such as

$$V = V_i(1 + \rho_v) \tag{4-42}$$

$$i = I_i(1 - \rho_v) \tag{4-43}$$

where ρ_v is the voltage reflection coefficient given in Eq.(4-27). Using Eqs.(4-42) and (4-43), one can carry out the analysis and evaluation of the wave behavior along the line more easily, because for a pulse travels with a constant speed, the time delay, the reflection coefficient and the locations of reflection are generally known. The only rules that need to be kept in mind are that (1) the polarities of the reflected waves must be properly observed as a voltage reflection is positive whereas a current reflection is negative. (2) at any given point along a line, the voltage or current at any given instant of time is the algebraic sum of the positively and negatively going waves. For example, the total voltage of two opposite going voltage waves of the same polarity and equal magnitude at a given point and time is equal to twice the individual wave magnitude. Similarly, the total voltage or current at the point of reflection, whether due to improper termination or discontinuity, is the algebraic sum of the incident and reflected waves. Let us use the following example to illustrate how these rules can be applied. Fig.4-7 shows an ideal transmission line terminated with a resistance R_l where $R_l < Z_0$. The input voltage is a step function of zero risetime, i.e., $V = V_s$ for $t \geq 0$ and the impedance of the voltage source is $R_s = Z_0$. Assuming at $t = 0$, the switch closes, we like to know what are the respective waveforms of the voltage and current at the input ($x = 0$) and output ($x = l$) ends. From Eq.(4-27) we have $\rho_v = (R_l - Z_0)/(R_l + Z_0) < 0$ and from Eq.(4-19) we know that for $0 \leq t \leq 2T_0$, the voltage at $x = 0$, is $V_i = V_s/2$ and the current is $I_i = V_0/Z_0 = V_s/2Z_0$. For $t > 2T_0$, the reflected wave V_r arrives at the input end, hence the total voltage at $x = 0$ is the sum of the initial and reflected voltages i.e.

$$V_0 = V_i(1 + \rho_v) = \frac{1}{2}V_s\left(1 + \frac{R_l - Z_l}{R_l + Z_l}\right) = \frac{V_s R_l}{R_l + Z_l}$$

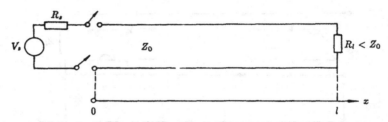

Fig.4-7 An ideal transmission line terminated with $R_s = Z_0$ and $R_l < Z_0$.

and the total current is

$$i_0 = V_i(1 - \rho_v) = \frac{V_s}{2Z_0}\left(\frac{2Z_0}{R_l + Z_0}\right) = \frac{V_s}{R_l + Z_0}$$

At the output end i.e. $x = l$, for $t < T_0$, nothing initiated from the input end has arrived yet, therefore both $V_l = 0$ and $i_l = 0$. For $t \geq T_0$, the total voltage is the sum of the incident and reflected voltages, namely

$$V_l = V_i(1 + \rho_v) = \frac{V_s R_l}{R_l + Z_l}$$

and similarly the current is

$$i_l = I_i(1 - \rho_v) = \frac{V_s}{R_l + Z_0}$$

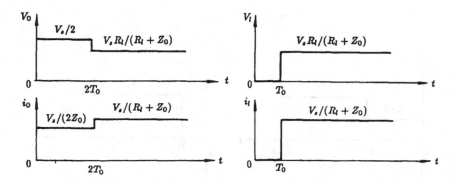

Fig.4-8 Step function response of an ideal line with $R_s = Z_0$ and $R_l < Z_0$, $T_0 = l\sqrt{LC}$ (see Fig.4-7).

The overall result is shown in Fig.4-8. From the figure, one can see that after $t = 2T_0$, the voltage and current become steady along the entire line. The respective waveforms as a function of time appearing at the input and output ends of the same line shown in Fig.4-7 for various other terminations are shown in Fig.4-9.

In the above example, the ideal line was assumed to be improperly terminated only at the output end ($R_l \neq Z_0$), so the waves along the line suffer only one reflection at the output end. If both the input and output ends are improperly terminated ($R_s \neq Z_0$ and $R_l \neq Z_0$), then multiple reflections at both ends will take place. Let us assume that the source impedance is $R_s < Z_0$, and the rest of the conditions in Fig.4-7 remain unchanged. In this case, we have two reflection coefficients ρ_0 and ρ_l, one for the input end and the other for the output end. From Eq.(4-27), we have

$$\rho_0 = \frac{R_s - Z_0}{R_s + Z_0} \quad \text{and} \quad \rho_l = \frac{R_l - Z_0}{R_l + Z_0}$$

and both are negative values. When the switch in Fig.4-7 is closed at $t = 0$, a step function voltage $V_i = V_s Z_0/(R_s + Z_0)$ appears at $x = 0$ on the line and propagates toward the load end ($x = l$). When it reaches the load at $t = T_0$, the first load reflection takes place and with a reflected wave of $\rho_l V_i$. This reflected wave travels toward the voltage source. When it reaches the input end at $t = 2T_0$, the first reflection at the voltage source occurs and with a value of $\rho_0 \left(\rho_l V_i\right)$. This reflected wave, in turn, travels toward the load where it is reflected, travels back to the input end where it is once again reflected. Such back and forth reflection continues until the reflected wave finally approaches zero. These various reflections and the times at which they occur are summarized in Table 4-1. It should be noted that here we

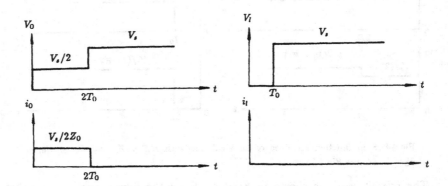

Fig.4-9(a) Step function response of an ideal line (Fig.4-7) with $R_s = Z_0$, $R_l = \infty$ and $T_0 = l\sqrt{LC}$.

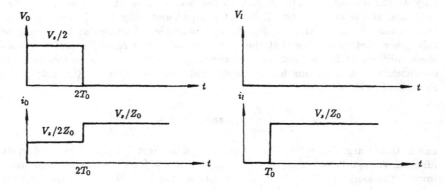

Fig.4-9(b) Step function response of an ideal line (Fig.4-7) with $R_s = Z_0$, $R_l = 0$ and $T_0 = l\sqrt{LC}$.

Table 4-1 Time of Occurrence and Amplitude of Reflected Waves in Fig.4-7

Reflection sequence	Reflection at load		Reflection at input	
	Time	Amplitude	Time	Amplitude
1st	T_0	$\rho_l V_i$	$2T_0$	$\rho_0 \rho_l V_i$
2nd	$3T_0$	$\rho_l(\rho_0\rho_l V_i)$	$4T_0$	$\rho_0(\rho_0\rho_l^2 V_i)$
3rd	$5T_0$	$\rho_l(\rho_0^2\rho_l^2)V_i$	$6T_0$	$\rho_0(\rho_0^2\rho_l^3 V_i)$
4th	$7T_0$	$\rho_l(\rho_0^3\rho_l^3)V_i$	$8T_0$	$\rho_0(\rho_0^3\rho_l^4 V_i)$

are concerned with the individual reflected waves (V_r in Eq.(4-38)) only, and not with the total observable voltage on the line. The total voltage V at any point along the line is a sum of the existing voltage at that point immediately prior to the reflection taking place and the incident and reflected voltages at the same point. For example, at the input end at $t = 4T_0$, a reflection occurs for which the incident wave is $\rho_l(\rho_0\rho_l V_i)$, the reflected wave is $\rho_0(\rho_0\rho_l^2 V_i)$, the existing wave immediately prior to $t = 4T_0$ is $(V_i + \rho_l V_i + \rho_0\rho_l V_i)$, and the total voltage is the sum of the three. The general expression for the total voltage at the input end is

$$V_0 = V_i \left\{ 1 + \sum_{\nu=1}^{k} [\rho_l^\nu \rho_0^{\nu-1} + \rho_l^\nu \rho_0^\nu] \right\} \tag{4-44}$$

where k is the number of round trips, and is equal to the reflection number shown in Table 4-1. At the output end, the total voltage is

$$V_l = V_i \left\{ 1 + \sum_{p=1}^{n} \rho_l^p \rho_0^{p-1} + \sum_{p=1}^{(n-1)} \rho_l^p \rho_0^p \right\} \tag{4-45}$$

where n is the number of times the wave has reached the load and is equal to the reflection number in Table 4-1. For $n < 2$, the terms in the 2nd summation in Eq.4-45 is equal to zero. In Eqs.(4-44) and (4-45), the terms in the first summation sign represent reflections originating at the load end while the terms in the second summation represent reflections occurring at the input end.

In a similar manner, the corresponding expressions for the current can be obtained which are identical to that given in Eqs.(4-44) and (4-45) expect that V_i in these equations is replaced by

$$I_i = \frac{V_s}{R_s + Z_0}$$

and change the sign of the voltage reflection coefficient from positive to negative, because the current reflection coefficient is the negative of the former. For this reason, as the amplitude of the total voltage decreases with time, the total current will build up, and vice versa. From the expressions given in Eqs.(4-44), (4-45) and the corresponding ones for the current, the total voltage and current appearing at the input and output ends of an ideal transmission line shown in Fig.4-7 can be readily obtained. The results for $R_s = Z_0/2$ and $R_l = Z_0/3$ are shown in Fig.4-10. From these results one can see that at the input and output ends on the line, the waveforms of the total voltage are not identical, and there is a time delay T_0 between the two. After a sufficiently long time, the voltages along the entire line will approach a steady value of $V_s[R_l/(R_l + R_s)]$. Similar situation exists for the current, except the current increases with time whereas the voltage V_0 decreases as can be seen from the figure.

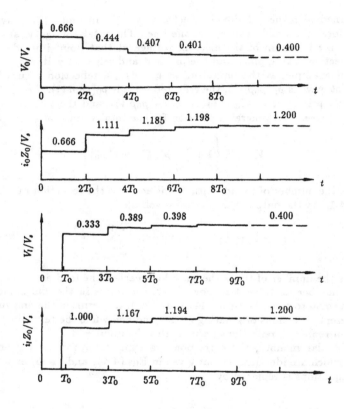

Fig.4-10 Step function response of an ideal line (Fig.4-7) improperly terminated at both ends with $R_s = Z_0/2$, $R_l = Z_0/3$ and $T_0 = l\sqrt{LC}$.

(b) Step Function Input With Linear Rise Time

In the previous sections, we have discussed only the cases involving the step functions of zero risetime. In practice, all step functions have a finite measurable risetime and quite often the risetime is comparable to the delay time of the line. In such cases, the waveforms will be greatly influenced by the risetime of the step function. Therefore proper treatment of the risetime of the input is necessary. For simplicity, let us assume that the applied voltage is a step function with a linear risetime as shown in Fig.4-11(a). If the transmission line is improperly terminated only at one end, e.g. $R_l < Z_0$, the situation is relatively easy to be dealt with as the wave will be reflected only once along the line. In this case, one can follow the same procedure described in Section 4-2(a) for the case of step function with zero risetime to deal with the problem except taking the risetime into consideration. Since the only reflection is at the load end, after the switch is closed, at $t = t_r + 2T_0$, the voltage along the line will become steady. The steady value of the voltage is $V_s[R_l/(R_l + Z_0)]$ which may be obtained from Eq.(4-42) by adding up the initial input voltage and the reflected one at the load. The waveform between $t = 0$ and $t = t_r + 4T_0$ may be obtained from a geometrical construction of the applied and reflected voltages superimposed with the proper time sequence. For example, when the risetime t_r is less than twice the delay time of the line $(2T_0)$, the reflected wave will arrive at the input end at an instant of $(2T_0 - t_r)$ after the applied voltage has reached its peak value as shown in Fig.4-11(b). If the risetime t_r equals just twice the delay time of the line, the reflected voltage arrives at the input end the instant the applied voltage has reached peak value. The result is a sharp peak of value $V_s/2$ as shown in Fig.4-11(c). If the delay time of the line becomes less than the risetime, the reflected voltage arrives at the input end before the applied voltage has reached peak value. The reflected voltage will tend to decrease while the applied voltage tends to increase the total voltage at the input end. The result is a waveform with three distant slopes as shown in Fig.4-11(d). From the figure, one can see that the peak value in this case is less than $V_s/2$ whereas the final value is identical to that of the previous cases.

If the transmission line is improperly terminated at both ends then multiple reflections will be present along the line. To deal with such a situation, one needs to pay more attention to the relationship between the various reflections at the proper time. In addition one also needs to apply the relations given in Eqs.(4-44) and (4-45) instead of that given in Eqs.(4-42) and (4-43) as the latter can only be applied directly to the cases involving one reflection. More specifically, the essential steps needed to be taken in this case are, firstly to find the initial input voltage and the reflection coefficients from the given data, secondly to determine the various reflected components at the proper times and locations, finally to employ the relations given in Eqs.(4-44) or (4-45) to obtain the total voltage at the input end or the output end. In general, the situation may be classified into three categories: (1) the delay time of the line T_0 is much larger than the risetime t_r. (2) the delay time is considerably shorter than the risetime. (3) the delay time

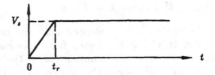

(a) Step function with linear rise time t_r

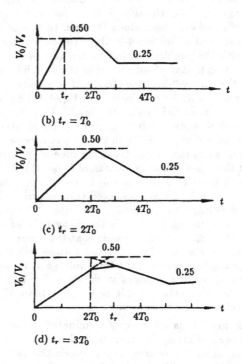

(b) $t_r = T_0$

(c) $t_r = 2T_0$

(d) $t_r = 3T_0$

Fig.4-11 Response of an ideal line (Fig.4-7) improperly terminated at one end $(R_s = Z_0, R_l = Z_0/3)$. Input is a step function with a linear risetime t_r.

is comparable to the risetime. For the first category, the rise-time may be taken as zero and the result of it is essentially identical to that shown in Fig.4-10. We shall not repeat the discussion here. For the second category, let us consider the case with $t_r = 4T_0$, $R_s = 2Z_0$ and $R_l = 3Z_0$, as shown in Fig.4-12. From Eq.(4-27), we know that the voltage reflection coefficient $\rho_0 = 1/3$ and $\rho_l = 1/2$. The initial voltage at the input end is $V_i = I_i Z_0$. With the peak current of $\bar{I}_i = V_s/(2Z_0 + Z_0)$, we have the peak voltage at the input end $\bar{V}_i = V_s/3$. This peak value of the voltage can be reached only if there is no reflection coming from the load end. In reality, however, at proper times there are various incident and reflected waves present at the input end. The total voltage is the sum of these voltages plus the initial one. The first incident and reflected voltages appearing at the input end is

$$V_1 = \rho_l V_i + \rho_0(\rho_l V_i) = 2V_i/9$$

The first term in the expression represents the reflected voltage from the load and the second term is the reflection at the input end. The reflected voltage $\rho_0(\rho_l V_i)$ at the input will be again reflected at the load end when it reaches there. Upon returning to the input end it will act as an incident wave to the input end, meanwhile it will produce another reflection. The sum of the two is

$$V_2 = \rho_0 \rho_l^2 V_i + \rho_0^2 \rho_l^2 V_i = V_i/(27)$$

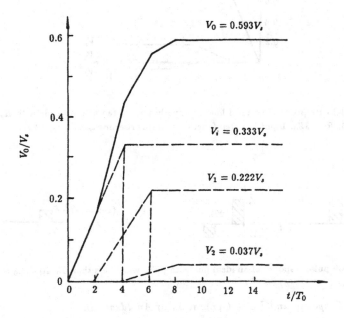

Fig.4-12 Response of a transmission line improperly terminated at both ends with $R_s = 2Z_0$, $R_l = 3Z_0$ and input is a step function of linear risetime t_r with $t_r = 4T_0$.

The voltage expressed by the second term will be further reflected at the load end and will produce more reflections, however, the magnitudes of the reflected waves become increasingly smaller and smaller. In practice the total voltage at the input end can be approximated as

$$V_0 \sim V_i + V_1 + V_2 = V_i(1 + \rho_l + \rho_0\rho_l + \rho_0\rho_l^2 + \rho_0^2\rho_l^2)$$

By substituting the peak value of V_i into the above expression, one can obtain the peak value of $\bar{V}_0 = 16V_s/27$. The various waveforms at the input end so obtained are shown in Fig.4-12. Fig.4-13 is the result obtained by the same procedure for the case with $t_r = 2T_0$, $R_s = Z_0/3$ and $R_l = 3Z_0$. As one can see from the figure, in this case the total voltage at the input end oscillates about the value $9V_s/10$ as it approaches the final value.

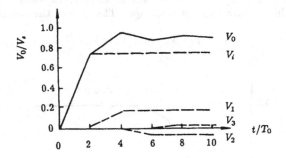

Fig.4-13 Response of an ideal line improperly terminated at both ends with $R_s = Z_0/3$, $R_l = 3Z_0$. Input is a step function of linear risetime t_r with $t_r = 2T_0$.

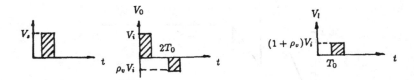

Fig.4-14 Square pulse response of an ideal line with pulse width less than T_0, $R_l < Z_0$ and $R_s = Z_0$.

(c) Pulse Response and Pulse Generation of An Ideal Line

In the proceeding sections, we have discussed the various aspects concerning the waveforms obtained with a step function input. A step function can be considered as a pulse of infinite width. When the width of the pulse is finite and shorter than the delay time of the transmission line, the individual reflections of the pulse can be seen separately at the input end, even the input end is terminated properly.

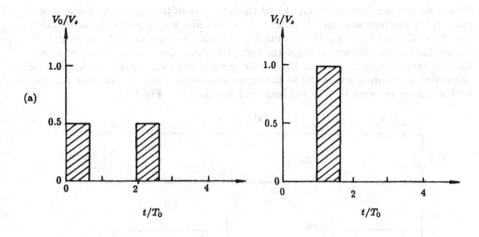

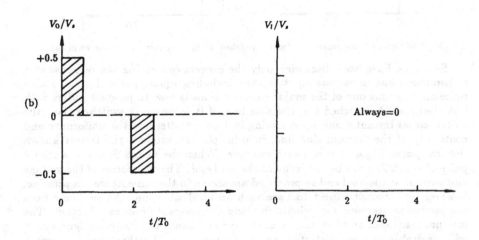

Fig.4-15 Square pulse response of an ideal line with pulse width less than T_0, $R_s = Z_0$ and (a) $R_l = \infty$, (b) $R_l = 0$ at the load end.

For example, in Fig.4-7, if the input voltage is a square pulse having a width less than the delay time T_0 of the transmission line, the waveforms of the voltage at the input and output ends would appear as that shown in Fig.4-14. From the figure one can see that at the input end the reflected voltage appears as a separate pulse from the incident one. If both ends of the line are improperly terminated, e.g. $R_s < Z_0$ and $R_l < Z_0$, then multiple reflections will occur and the result will be similar to that shown in Fig.4-10, except in this case, the various reflections will appear as separate pulses. Under the same conditions mentioned above, the respective waveforms appearing at the input and output ends of the same line but with different terminations at the load end are shown in Fig.4-15.

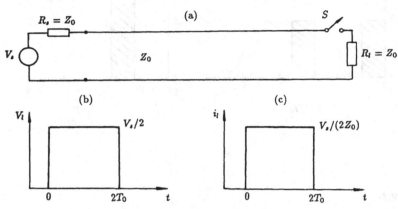

Fig.4-16 Square pulse generation from a matched, ideal transmission line shown in (a).

So far we have been discussing only the aspects concerning the response of a transmission line to various input voltage, including square pulse. In high power pulse applications one of the major considerations is how to produce pulses with fast risetime. A matched transmission line, if it is properly switched, i.e. the switch closes instantly and upon closing it does not alternate the uniformity and continuity of the transmission line, in principle can produce the fastest known risetime pulse. Fig.4-16 is one such example. When the switch S closes, a square pulse of width $2T_0$ will be produced at the load end. The waveforms of the voltage and current at the load end so produced are shown in the same figure. In practice, however, one cannot expect to have such an ideal situation. As we know from the proceding sections, any mismatch along the line will produce reflection. The internal resistance and inductance of the switch cannot be altogether ignored. It will inevitably produce distortion on the output pulse. Furthermore, the switch can never be closed instantaneously. In other words, the risetime and falling time of the switch can also have a great effect on the waveform of the output pulse. Therefore the actual form of the pulse produced from the set-up shown in Fig.4-16(a) can be expected more like that shown in Fig.4-17 rather than the ones shown in Fig.4-16(b) and (c). For these reasons, the switching technique is one important area to be concerned with in high power pulse applications. In a later chapter, we shall give a detailed discussion on the subject.

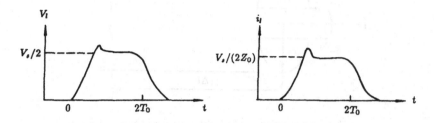

Fig.4-17 Actual pulse form can be expected from the matched transmission line shown in Fig.4-16(a).

4-3 Parameters of Transmission Lines of Different Geometrical Configurations

In the previous sections we mainly considered the fundamental principles and theoretical formulations which govern the basic behavior of the transmission line. The parameters R, G, L and C which are required to construct the equivalent circuit of the transmission line, were only discussed in the general sense and no specific details were given to evaluate these parameters for any given conditions, such as for a given geometrical configuration of the transmission line. In this section, we shall, from the practical point of view, discuss how to evaluate these parameters for various geometrical configurations of practical interest and present related data as useful reference. The other purpose of this section is to pull together the essential information concerning ideal lines of various geometrical configurations into a compact form so that the readers can easily get these information for practical applications. Since we are dealing with ideal lines, the parameters R and G in the following discussions shall be taken as zero and only the parameters L and C are to be considered.

(a) Coaxial Cylindrical Line

Let us consider a uniform line of infinite length so that the fields are uniform everywhere and the fringing effect is negligible. Assuming that the inner and outer radius of the cylinders are r_1 and r_2, respectively, the voltage between the two conductors is V, and the permittivity and permeability of the medium between the two conductors are respectively ϵ and μ as shown in Fig.4-18. The inductance per unit length of the line in general is frequency dependent. At dc or low frequency, the current is uniformly distributed over the conductor cross-section areas, the inductance may be expressed as consisting of two parts, one arising from the magnetic flux linkages in the region between the two conductors and the other as a result of magnetic flux linkages within the conductors. Mathematically it can be

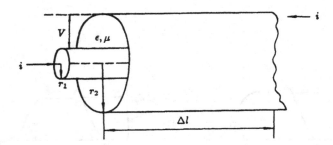

Fig.4-18 A coaxial cylindrical line with voltage and current indicated.

expressed as

$$L = \left\{ \frac{\mu}{2\pi} \ln \frac{r_2}{r_1} + \frac{\mu}{8\pi} \right\} \quad \text{(h/m)} \tag{4-46}$$

At sufficiently high frequencies, the skin effect causes the current to flow only on the outer surface of the inner conductor and the inner surface of the outer conductor. As a consequence, the second term in Eq.(4-46) disappears and the inductance per unit length then becomes

$$L = \frac{\mu}{2\pi} \ln \frac{r_2}{r_1} \quad \text{(h/m)} \tag{4-47}$$

or

$$L = 0.2\mu_r \ln(r_2/r_1) \quad (\mu\text{h/m})$$

where μ_r is the relative permeability. The capacitance per unit length is basically frequency independent and can be expressed as

$$C = \frac{2\pi\epsilon}{\ln(r_2/r_1)} \quad \text{(F/m)} \tag{4-48(a)}$$

or

$$C = \frac{55.6\epsilon_r}{\ln(r_2/r_1)} \quad (\mu\mu\text{F/m}) \tag{4-48(b)}$$

where $\epsilon = \epsilon_0 \epsilon_r$ and ϵ_r is the relative permittivity. Substituting Eqs.(4-47) and (4-48) into Eq.(4-30) and ignoring R and G, we obtain the characteristic impedance of the line

$$Z_0 \simeq 60 \left(\frac{\mu_r}{\epsilon_r} \right)^{1/2} \ln \frac{r_2}{r_1} \quad (\Omega) \tag{4-49}$$

which indicates that for a given medium between the conductors, the characteristic impedance of a coaxial line is determined by the ratio of the radius of the two conductors. This is an important feature of the transmission line which enables one to construct a transmission line of various sizes with the same characteristic impedance.

(b) Parallel Cylindrical Line

When two infinitely long cylindrical conductors of the same diameter are arranged in a parallel configuration as shown in Fig.4-19, the inductance per unit length can be shown to be[4],

$$L = \frac{\mu}{\pi} \ln \left\{ \frac{d + (d^2 - 4r^2)^{1/2}}{2r} \right\} \quad \text{(h/m)} \quad (4\text{-}50)$$

and the capacitance per unit length is

$$C = \frac{\epsilon \mu}{L} = \frac{\pi \epsilon}{\ln \left[\dfrac{d + (d^2 - 4r^2)^{1/2}}{2r} \right]} \quad \text{(F/m)} \quad (4\text{-}51)$$

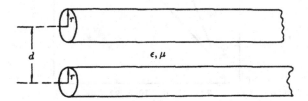

Fig.4-19 A parallel cylindrical line.

The characteristic impedance of the line can be readily obtained as

$$Z_0 = 120 \left(\frac{\mu_r}{\epsilon_r} \right)^{1/2} \ln \left[\frac{d + (d^2 - 4r^2)^{1/2}}{2r} \right] \quad (\Omega) \quad (4\text{-}52)$$

When the separation of the two cylinders is very large compared to the radius of the cylinders, Z_0 can be approximated as

$$Z_0 \simeq 120 \left(\frac{\mu_r}{\epsilon_r} \right)^{1/2} \ln \frac{d}{r} \quad (\Omega) \quad (4\text{-}53)$$

which is similar to the expression given in Eq.(4-49) for coaxial cylindrical line. The application of this type of transmission line is common in the conventional power transmission systems. In high power pulse systems, the coaxial type is usually employed when a cylindrical line is required. Fig.4-20 is a chart showing the relationships between the normalized impedance and the typical values of r_2/r_1 for coaxial line and d/r for parallel line.

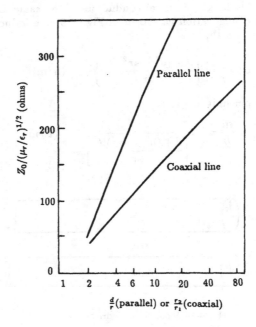

Fig.4-20 Characteristic impedance versus geometrical ratio for coaxial and parallel cylindrical lines.

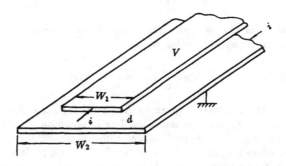

Fig.4-21 A general strip line.

(c) Strip Line

Strip line is one of the widely used devices in high power pulse applications. Proper understanding of the applicability and limitation of the parameter relationships is important when the design of such a transmission line is attempted. A general strip line is usually defined as two parallel plates of different widths W_1 and W_2, separated by a fixed distance d as shown in Fig.4-21. Analytic expressions of the inductance and capacitance in terms of W_1, W_2 and d cannot be readily obtained. For an infinitely long line, the characteristic impedance can be expressed in terms of the capacitance C as

$$Z_0 = \frac{3.33 \times 10^{-9}(\mu_r \epsilon_r)^{1/2}}{C} \quad (\Omega) \qquad (4\text{-}54)$$

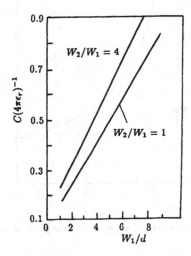

Fig.4-22 Capacitance versus geometrical ratio for strip lines having different values of W_2/W_1 [5].

where C is the capacitance per unit length and expressed in F/M. It is a complex function of W_1, W_2 and d and difficult to be directly evaluated[5]. The values of C in practice are usually tabulated in tables or plotted as curves. Fig.4-22 is one of them. Fig.4-23 is a plot of Z_0 versus W_1/d. When the widths of the two conductors are identical ($W_1 = W_2$) and the ratio of W/d is sufficiently large (10 or larger), the parameters of the line can be expressed analytically. They are

$$L = 4\pi \times 10^{-7} \mu_r \frac{d}{W} \quad (\text{h/m}) \qquad (4\text{-}55)$$

$$C = 8.85\epsilon_r \frac{W}{d} \qquad (\mu\mu\text{F/m}) \qquad (4\text{-}56)$$

and

$$Z_0 = \sqrt{L/C} = 377\frac{d}{W}\left(\frac{\mu_r}{\epsilon_r}\right)^{1/2} \quad (\Omega) \qquad (4\text{-}57)$$

From Eq.(4-57) one can see that the characteristic impedance of an equal-width strip line is linearly dependent on the ratio of d/W whereas in the cylindrical transmission line, it is dependent on the logarithm of r_1/r_2 or d/r.

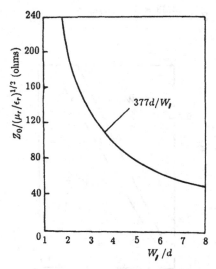

Fig.4-23 A plot of the characteristic impedance versus geometrical ratio for a general strip line.

(d) Helical Line

In high power pulse applications, it is frequently required to employ delay line as timing, synchronization etc. Quite often, the required delay times are very large, and using an ordinary transmission line becomes impractical. One way to increase the delay time without physically increasing the length of the line is to construct the inner conductor with helical winding as shown in Fig.4-24 rather than straight. In this case two kinds of dielectric mediums ϵ_1, μ_1 and ϵ_2, μ_2 are generally employed, one is used between the two cylinders and the other is for the helical winding as shown in the figure. The determination of the inductance and capacitance of such a line is mathematically quite complicated[6]. We shall give

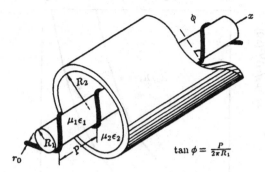

$$\tan \phi = \frac{p}{2\pi R_1}$$

Fig.4-24 A helical transmission line.

the end result only. The general expression of the characteristic impedance is

$$Z_0 = \frac{R_1}{p} \left(\frac{\mu_r}{\epsilon_r}\right)^{1/2} (S_1 S_2)^{1/2} \qquad (\Omega) \qquad (4\text{-}58)$$

and the phase velocity in the axial direction

$$v_x = \frac{\tan \phi}{(\mu_r \epsilon_r)^{1/2}} \left(\frac{S_1}{S_2}\right)^{1/2} \qquad (\text{m/s}) \qquad (4\text{-}59)$$

where S_1 and S_2 are functions of r_0, R_1, R_2, ϵ_1, μ_1, ϵ_2 and μ_2. For the special cases of $\epsilon_1 = \epsilon_2$, $\mu_1 = \mu_2$ and $R_1 \gg r_0$, the impedance becomes

$$Z_0 = \frac{1}{2\pi} (\mu/\epsilon)^{1/2} \ln \left[\frac{R_2}{R_1} - \frac{R_1}{R_2}\right] \qquad (\Omega) \qquad (4\text{-}60)$$

From Eq.(4-59), one can determine the delay time. Curves for the impedance and phase velocity for various geometrical ratio are shown in Fig.4-25 and Fig.4-26 where the entire line was assumed to have a dielectric constant ϵ_2 and with a very large value of μ_1/μ_2. As would be expected, the phase velocity in the axial direction decreases or the delay time increases rapidly as the turns become more closely spaced. Meanwhile the impedance of the line also increases and a compromise between these two is often necessary.

(e) Tapered Line

In some applications, the characteristic impedance of a transmission line is required to vary gradually along the length so that it can properly join two devices of different characteristic impedance. In some other applications, in order to effectively transport the power from a device of large size to a target of smaller size the characteristic impedance of the line is required to be constant while its physical size changes gradually along the length so that it can match the two devices of

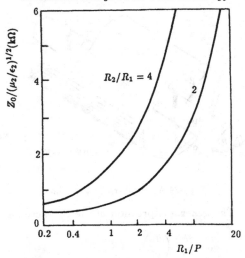

Fig.4-25 Characteristic impedance versus geometrical ratio for a helical line with μ_1/μ_2 very large, $\epsilon_1 \simeq \epsilon_2$ and $r_0/R_1 = 0.01$.

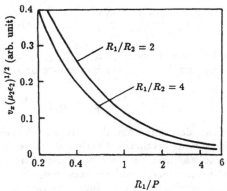

Fig.4-26 Axial phase velocity versus geometrical ratio for a helical line with $\mu_1 \gg \mu_2$, $\epsilon_1 \simeq \epsilon_2$, and $r_0/R_1 = 0.01$.

different sizes properly. In both cases, one can employ the tapered transmission line to achieve the objective. For the first case, one conductor of the transmission line is made tapered along its length while the other conductor is kept uniform as shown in Fig.4-27(a). The respective impedance at the two ends of the line is

$$Z_1 = 60 \left(\frac{\mu_r}{\epsilon_r}\right)^{1/2} \ln \frac{r_2}{r_1} \quad (\Omega)$$

$$Z_2 = 60 \left(\frac{\mu_r}{\epsilon_r}\right)^{1/2} \ln \frac{r_2}{r_1'} \quad (\Omega)$$

In the second case, the two conductors of the line are made both tapered along the length meanwhile the ratio of their radii r_2/r_1 is kept constant as shown in Fig.4-27(b). From the above expressions, it is obvious that the characteristic impedance Z_0 in this case is a constant inspite of its size changes along the length. The above illustrations are two typical examples of the applications of the tapered transmission line in high power pulse application. The tapered transmission line can also be employed in a number of other applications[7]. Further discussion on the subject shall be given in a latter section. In the above example, coaxial cylindrical lines were employed to demonstrate the essential properties of the tapered line, which by no means implies that the only coaxial line is suitable for such purpose. In fact, all other types of transmission lines mentioned above are equally suitable to be constructed as a tapered transmission line.

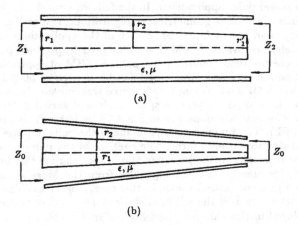

Fig.4-27 Tapered transmission lines. (a) constant outer diameter variable characteristic impedance; (b)constant characteristic impedance variable diameters.

When the above formulas, Eqs.(4-46) to (4-60) are employed for practical applications, one should bear in mind that there are considerable limitations imposed on them. Strictly speaking, these formulas are correct only for the ideal cases, namely an infinitely long (or properly terminated), lossless and dispersionless transmission line. In practice, as we know, most of the conditions are far from ideal, therefore one should not expect high accuracy in the results when those formulas are employed.

4-4 Simple Transmission Line

(a) Some Typical Examples

Any transmission line which is structured with two pieces of conductors may be called a simple transmission line. The transmission lines used in the previous illustrations, including those shown in Section 4-3, are all simple transmission lines. Nearly all the fundamental formulations concerning transmission lines were derived on the basis of a simple transmission line. Therefore, understanding of simple transmission lines is essential and important for the understanding and operation of other types of transmission lines, as all other types of transmission lines (excluding MITL) are developed on the basis of the former. As the name implies, simple transmission lines are simple to construct and easy to operate as compared with other types of transmission lines. The main drawback is that it is incapable of yielding output voltage more than half of the charging voltage. Inspite of such shortcomings, a simple transmission line has been widely employed in various high power pulse applications. In the following we shall use three simple transmission lines of different geometrical configurations, respectively to illustrate what role it plays and how it functions in each of the applications. In Fig.4-28 is an electron accelerator system[8] which was designed to produce pulsed electron beam currents greater than 500kA with a pulse FWHM of approximately 65 ns. The electron beam energy and power level can be as high as 50 KJ and 0.7 TW respectively when a 5Ω diode is employed. Three transmission lines are employed in this system. Each of them plays a special role and serves different functions in the system. The first one indicated in Fig.4-29 by letter A, is used as a pulse forming line (PFL). The function of it is to shape the voltage pulse coming from the Marx generator to a proper form. The risetime of the pulse produced from a Marx generator is usually in the μs range and is too long for this purpose. With a PFL properly designed, the incoming pulse from the Marx generator can be re-shaped into a pulse of desired width, in this case, it is 77 ns. The peak voltage in the PFL is determined by the self-breakdown parameters of the output switch. The PFL employed in this case has the following specifications:

Diameter of the outer conductor	122 cm
Diameter of the inner conductor	58 cm
Length of the PFL	130 cm

With the dielectric constant for water = 80 substituting these figures into Eq. (4-49) and taking $\mu = 1$, one can find that the characteristic impedance of the PFL is

$$Z_0 = \frac{60}{\sqrt{\epsilon_r}} \ln \frac{r_2}{r_1} = 5\Omega$$

Further substituting the above figures into Eqs. (4-47), (4-48) and (4-41), one can also obtain the one way travelling time of the voltage wave along the PFL to be

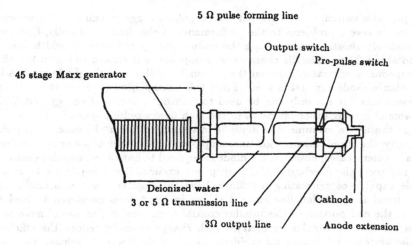

Fig.4-28 An electron accelerator system in which coaxial cylindrical T-L are employed [8].

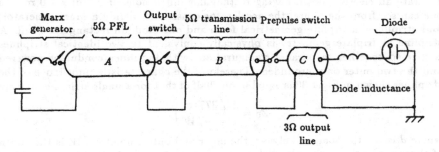

Fig.4-29 Schematic diagram of the electron accelerator system shown in Fig.4-28 [8].

approximately;

$$T_0 = 38.6 \quad \text{ns}$$

which gives a pulse width (FWHM) of approximately 77 ns as mentioned previously.

The second line indicated by B is a transmission line to transmit the electric power from the PFL to the load. It is charged by the output of the PFL when the output switch closes. In order to properly match the impedance of the PFL it has the same parameters as that of the former. The 130 cm length is chosen so that the pulse reflection from the diode and the prepulse switch can be delayed until after the main pulse begins to turn off. The third line indicated by C is a short line of 46cm length and is employed to serve two purposes. Firstly, by means of the prepulse switch, it is to prevent the prepulse voltage reaching the diode which can cause severe problems to the performance of the diode. Secondly, because it is relatively short, it can reshape the main pulse by reducing its width from 77 ns to about 27 ns. The 3Ω characteristic impedance is chosen to lessen the effect of impedance mismatch between the transmission line and the diode when a low impedance diode e.g. 1Ω is used. From this example, one can see that a simple transmission line not only can be used to transmit power and energy, but is also capable of serving other functions in the high power pulse applications.

In the above example, the diode is in cylindrical form. In order to match it properly, the three transmission lines employed were chosen to be coaxial cylinder lines. In some cases, however, the diode is required to have a rectangular geometry. For instance, the development of high power excimer lasers requires a large area diode capable of generating a 0.3m x 1.0m rectangular electron beam[9]. If a cylindrical transmission line as that shown in Fig.4-28 is employed to feed the diode, the end portion of the tubular coaxial transmission line would have to be changed into rectangular geometry. Such changes usually reduce the efficiency of the system and should be avoided. One practical way to achieve this is to employ a stripline instead of the coaxial one in the system. Fig.4-30 shows the basic structure of a 0.5 TW electron beam accelerator which was designed to generate an electron beam having rectangular dimensions of 0.3 m x 1.0 m. As one can see from the figure, the accelerator consists of mainly a Marx generator, four striplines in triplate geometrical form and one diode in rectangular form. A stripline in triplate geometry is physically equivalent to two identical striplines connected together in parallel configuration, with the inner conductor powered and the two outer ones grounded. Therefore, the characteristic impedance of the striplines shown in Fig.4-30 is equal to one half of that for a single stripline, namely

$$Z_0 = \frac{1}{2} \left(\frac{377}{\sqrt{\epsilon_r}} \frac{d}{W} \right)$$

where d is the the spacing between the inner and out conductors, W is the width of these conductors and ϵ the dielectric constant of water. In this case, $d = 0.1$m, $W = 1.07$m, and $\epsilon_r = 80$, substituting these values into the above equation, one will find that Z_0 is about 2Ω. It should be noted that the d for the first PFL (longer one) is 0.2m, which yields a higher impedance for this line. This is to have a better match with the intermediate store which has a higher impedance. The respective functions of the four striplines shown in Fig.4-30 are similar to that of the three coaxial lines described in the preceding example, except in this case there are two PFLs and the output line is named as convolute. Should the convolute be in cylindrical form, from the figure one can see that, it would require extensive change at its end to match the rectangular dimensions of the diode whereas with a stripline only a minor change is required.

To illustrate how a tapered transmission line functions in the high power pulse applications, we shall use an Electromagnetic Pulse (EMP) generator as an example. In a nuclear explosion, strong EMP is generated which is capable of impairing

Pulse forming lines   Oil/water interface

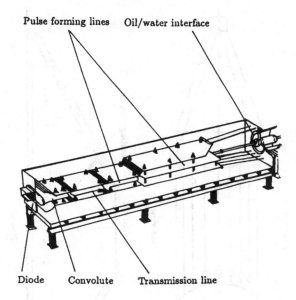

Diode   Convolute   Transmission line

Fig.4-30 Rayito accelerator in which tri-plate striplines are employed [9].

the performance of electrical and electronic systems. In order to study the effect a simulator capable of producing similar EMP is needed. The EMP required for such a purpose needs to have an electric field intensity greater than 10^5 v/m and risetime of 10 ns or less. EMP generator is the facility designed to meet these requirements. Fig.4-31 shows the essential structure of an EMP generator[10]. It consists of mainly a Marx generator as the energy source, two tapered transmission lines connected with wire antennas to generate the required EMP and a match load at the terminal to dissipate the EMP energy. The EMP test volume, i.e. where the test equipments are kept, is contained between the upper and lower electrodes of the stripline. The lower electrode is usually grounded. The dimensions of the test volume (or the transmission line)vary accordingly to the size of the objects to be tested. The height, or spacing, between the upper and lower electrodes can range from a few meters to tens of meters, the width and length can be tens of meters. To design a high power machine of this nature requires special consideration in several aspects. The output bushing of a typical Marx generator which is intended to supply power to the transmission line is normally coaxial with an outer diameter of one meter or less. A geometric transformation must be made between the coaxial bushing and parallel plate of the stripline so that they can join together smoothly without seriously degrading the pulse quality. To do so, the stripline is required to be tapered down as it approaches the Marx generator (or the load), and keeping the height to width ratio, or the impedance constant. However, as

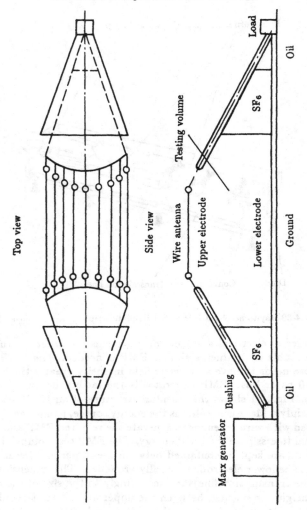

Fig.4-31 Basic structure of an EMP generator in which tapered striplines are employed.

the spacing between the upper and lower electrodes decreases toward the apex, at some point the spacing will not be adequate to support the peak voltage of the EMP and use of superior dielectric such as SF_6 or oil will become necessary. This in turn creates another problem of impedance mismatch at the interface between two different dielectrics. From this example, one can see that use of tapered transmission line in the present situation is technically more complex but all the objectives are achievable if one has sufficient knowledge in transmission line. For further information concerning the design of a lossless tapered transmission line, one may refer to Ref.[11].

(b) Circuit Analysis

When a Marx generator is employed as the energy source to feed a simple transmission line, the equivalent circuit of the system may be approximated by Fig.4-32. In the figure, R_g is the grounding resistance of the Marx generator, whose magnitude in general is much larger than that of R_l as illustrated in Fig.2-18, therefore can be neglected. The system, then can be represented by the circuit shown in Fig.2-10. In Section 2-3(b), we have carried out some preliminary analysis for the circuit and presented some basic results there, however, no details were given. In this section, we shall further carry out the analysis on the subject and give detailed procedures and results.

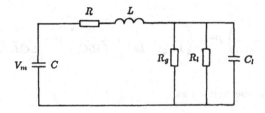

Fig.4-32 Approximate equivalent circuit of a Marx generator with a simple transmission line as its load, where V_m =maximum output voltage of the Marx generator, C =equivalent capacitance of the Marx generator, R =series resistance of the Marx generator, L =series inductance of the Marx generator, R_g =grounding resistance of the Marx generator (see Fig.2-18), R_l =equivalent shunt resistance of the transmission line, C_l =equivalent capacitance of the transmission line.

We begin the analysis with the basic equations given in Eq.(2-8) which were obtained from Fig.2-10 and based on Kirchhoff's law. These equations are linear equations with constant coefficients, and therefore, can be readily solved either by the usual way of solving linear differential equations or via the method of Laplace transformation. In this case, however, it is more convenient to employ the latter method. Upon Laplace transformation, the equations given in Eq.(2-8) become

$$\left.\begin{aligned} \frac{V_m}{P} &= \left[\frac{1}{CP} + LP + R\right] I(P) + R_l I_1(P) \\ I_2(P)/(C_l P) &- R_l I_1(P) = 0 \\ I(P) &= I_1(P) + I_2(P) \end{aligned}\right\} \tag{4-61}$$

where P is the new variable, $I(P)$, $I_1(P)$ and $I_2(P)$ are the corresponding functions of $i(t)$, $i_1(t)$ and $i_2(t)$ after Laplace transformation. By eliminating $I(P)$ and $I_1(P)$

from these equations, one can express $I_2(P)$ as

$$I_2(P) = \frac{V_m P}{L\left[P^3 + \left(\dfrac{R}{L} + \dfrac{1}{R_l C_l}\right)P^2 + \left(\dfrac{1}{LC_l} + \dfrac{1}{LC} + \dfrac{R}{LR_l C_l}\right)P + \dfrac{1}{LCR_l C_l}\right]}$$

(4-62)

Since the voltage across the transmission line is $V_{cl} = I_2(P)/(C_l P)$ so

$$V_{cl} = \frac{V_m}{LC_l}\left[P^3 + \left(\frac{R}{L} + \frac{1}{R_l C_l}\right)P^2 + \left(\frac{1}{LC_l} + \frac{1}{LC} + \frac{R}{LR_l C_l}\right)P + \frac{1}{LCR_l C_l}\right]^{-1}$$

(4-63)

Let λ_1, λ_2, and λ_3 be the three roots of the characteristic equation

$$P^3 + \left(\frac{R}{L} + \frac{1}{R_l C_l}\right)P^2 + \left(\frac{1}{LC_l} + \frac{1}{LC} + \frac{R}{LR_l C_l}\right)P + \frac{1}{LCR_l C_l} = 0 \qquad (4\text{-}64)$$

then Eq.(4-63) can be written as

$$V_{cl} = \frac{V_m}{LC_l(P - \lambda_1)(P - \lambda_2)(P - \lambda_3)} \qquad (4\text{-}65)$$

Upon inverse Laplace transformation of Eq.(4-65), the voltage across the transmission line as a function of time can be expressed as

$$V_l(t) = V_m \alpha^2 \left[\frac{e^{\lambda_1 t}}{(\lambda_1 - \lambda_2)(\lambda_1 - \lambda_3)} + \frac{e^{\lambda_2 t}}{(\lambda_2 - \lambda_1)(\lambda_2 - \lambda_3)} + \frac{e^{\lambda_3 t}}{(\lambda_3 - \lambda_1)(\lambda_3 - \lambda_2)}\right]$$

(4-66)

where $\alpha = 1/\sqrt{LC_l}$. Further from Eq.(4-64), one can obtain

$$\left.\begin{aligned}
\lambda_1 &= R + S - T/3 \\
\lambda_2 &= \frac{-1 + \sqrt{3}j}{2}R + \frac{-1 - \sqrt{3}j}{2}S - \frac{T}{3} \\
\lambda_3 &= \frac{-1 - \sqrt{3}j}{2}R + \frac{-1 + \sqrt{3}j}{2}S - \frac{T}{3}
\end{aligned}\right\} \qquad (4\text{-}67)$$

where

$$R = \left[-\frac{N}{2} + \sqrt{\left(\frac{N}{2}\right)^2 + \left(\frac{M}{3}\right)^3} \right]^{1/3}$$

$$S = \left[-\frac{N}{2} - \sqrt{\left(\frac{N}{2}\right)^2 + \left(\frac{M}{3}\right)^3} \right]^{1/3}$$

$$T = \frac{R_l C_l R + L}{L R_l C_l} \tag{4-68}$$

$$N = \frac{2R^3}{27L^3} + \frac{2}{27 R_l^3 C_l^3} - \frac{R}{3L^2 C_l} - \frac{R}{3L^2 C} - \frac{R^2}{9 L^2 R_l C_l}$$

$$\quad - \frac{R}{9 L R_l^2 C_l^2} - \frac{1}{3 L C_l^2 R_l} + \frac{2}{3 L R_l C_l C}$$

$$M = \frac{1}{L C_l} + \frac{1}{L C} + \frac{R}{L R_l C_l} - \frac{(R_l C_l R + L)^2}{3 L^2 R_l^2 C_l^2}$$

Substituting the above expressions into Eq.(4-66), one can obtain the explicit expression of the voltage across the transmission line in terms of time t. $V_l(t)$ can be either an oscillatory function of t or a non-oscillatory function of t depending on the magnitudes of the various parameters of the circuit. If $[(N/2)^2 + (M/3)^3] \leq 0$, the three roots λ_1, λ_2, and λ_3 are all real numbers and $V_l(t)$ is not oscillatory. When $[(N/2)^2 + (M/3)^3] > 0$, two of the three roots are complex numbers and $V_l(t)$ is then oscillatory. In high power pulse applications, the oscillatory mode is more useful than the non-oscillatory one[13] therefore, in this section we shall discuss the former only. The analytic expression of $V_l(t)$ in oscillatory mode is[12]

$$V_l(t) = V_m \alpha^2 \left\{ \frac{\exp(R + S - T/3)t}{3(R^2 + S^2 + RS)} + \frac{\exp(-R/2 - S/2 - T/3)t}{3(R-S)\sqrt{R^2 + S^2 + RS}} \right.$$

$$\left. \times 2\cos\left[\frac{\sqrt{3}}{2}(R-S)t - \left(k\pi + \mathrm{tg}^{-1}\frac{\sqrt{3}(R+S)}{(R-S)}\right) \right] \right\} \tag{4-69}$$

where $k = 0$ or 1

By differentiating Eq.(4-69) and using the relation $i_2(t) = C_l(dV_l/dt)$, one can obtain the time-dependent expression of the current going through the transmission line as

$$
i_2(t) = \frac{V_m}{L} \left\{ \frac{(R + S - T/3)\exp(R + S - T/3)t}{3(R^2 + S^2 + RS)} \right.
$$

$$
- \frac{\left(\dfrac{R}{2} + \dfrac{S}{2} + \dfrac{T}{3}\right)\exp\left(-\dfrac{R}{2} - \dfrac{S}{2} - \dfrac{T}{3}\right)t}{3(R - S)\sqrt{R^2 + S^2 + RS}}
$$

$$
\times 2\cos\left[\frac{\sqrt{3}}{2}(R - S)t - \left(k\pi + \text{tg}^{-1}\frac{\sqrt{3}(R + S)}{(R - S)}\right)\right]
$$

$$
- \frac{\sqrt{3}\exp\left(-\dfrac{R}{2} - \dfrac{S}{2} - \dfrac{T}{3}\right)t}{3\sqrt{R^2 + S^2 + RS}}
$$

$$
\left. \times \sin\left[\frac{\sqrt{3}}{2}(R - S) - \left(k\pi + \text{tg}^{-1}\frac{\sqrt{3}(R + S)}{(R - S)}\right)\right]\right\} \tag{4-70}
$$

With the expression given in Eqs.(4-69) and (4-70), one can further obtain other pertinent relations such as the power in the transmission line which is simply the product of V_l and i_2, and the energy stored in the transmission line which is $0.5C_lV_l^2$. By comparing these quantities with the corresponding ones of the Marx generator, one can evaluate the various efficiencies of the system. These include the power transfer efficiency η_P, which is defined as the ratio of the peak power in the transmission line to the peak power in the Marx generator, energy transfer efficiency η_ϵ and the voltage transfer efficiency η_v defined in the same way.

As an example, let us use the parameters of the Marx generator shown in Fig.2-17, namely $C = 0.02\mu$F, $L = 15\mu$h, $R = 5\Omega$, $V_m = -540$ kV and assume that the equivalent lumped capacitance and shunt resistance of the transmission line are $C_l = 0.02\mu$F and $R_l = 350\Omega$ respectively to carry out some calculations. From Eqs.(4-69) and (4-70) and other pertinent relations mentioned above, the various results so obtained are shown in Fig.4-33 and Table 4-2. Accompanying the calculated results are also the measured ones for comparison. From these results, one can see that the calculated results are generally consistent with the observed ones, however, discrepancy is also apparent. Discrepancy between them arises mainly from two major sources.

Table 4-2 Calculated and measured efficiencies of the Marx generator and transmission line system described in Fig.4-32.

Method	Energy transfer efficiency η_ϵ	Voltage transfer efficiency η_v
Calculated	72%	85%
Measured	69%	74%

They are (1) approximations made in setting up the equivalent circuit, (2) inaccuracy in the measurements. Discrepancy arising from the first source may

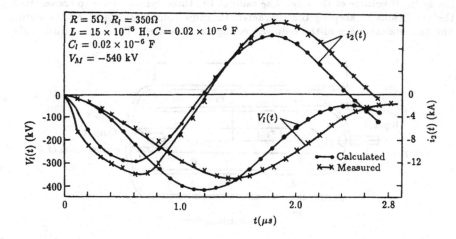

Fig.4-33 Calculated and measured results of $V_l(t)$ and $i_2(t)$ in a transmission line charged by a Marx generator as described in Fig.4-32.

be improved by using the numerical method to solve the governing equations. In this approach, a more accurate and realistic circuit, which normally is difficult to be dealt with analytically, may be employed to represent the system. Various numerical codes for this purpose have been developed in the past[14−17]. Proper use of such codes may improve the calculated results considerably. Discrepancy arising from the second source is rather circumstantial and largely determined by the measuring technique and instruments employed which may differ significantly from one case to another, and therefore is difficult to be discussed in the general sense. For this reason, we shall omit the discussion of it.

4-5 Blumlein Line

(a) Basic Principle

The basic structure of a Blumlein transmission line, as mentioned previously, may be conceptually considered as two simple transmission lines connected in a way that they are charged in parallel configuration and discharged in series. For this reason, it can, with proper termination, produce output voltage twice as high as that of a simple transmission line. Blumlein can be constructed either in cylindrical form or in parallel plate configuration. In cylindrical form, its basic structure consists of mainly three coaxial cylinders of different radii and they are arranged in a way shown in Fig.4-34. The space between the cylinders is filled with liquid dielectric medium, such as oil or water, for most high power pulse applications. Between the middle and inner cylinders there is a switch to control the hold-off voltage of the line. The radii of the three cylinders are so chosen that the conditions of keeping the characteristic impedance uniform along the entire line and maintenance of the required voltage are satisfied. The load is commonly

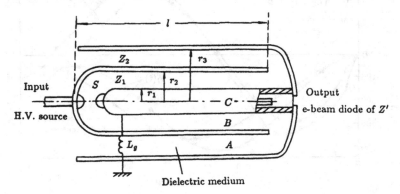

Fig.4-34 Basic structure of a cylindrical Blumlein line with an e-beam diode as the matched load where A=outer cylinder of radius r_3, B=middle cylinder of radius r_2, C=inner cylinder of radius r_1, S=H.V. switch, Z_1, Z_2 = characteristic impedances of the inner and outer lines, Z' = load impedance, L_g = grounding inductor.

connected between the inner and outer cylinders and the high voltage input is fed in via the middle cylinder as indicated in the figure. For the parallel plate configuration, its basic structure is shown in Fig.4-35. From the figure one can see that, other than the geometrical configuration, its basic elements and essential features are identical to that of a cylindrical Blumlein line, therefore we shall not repeat the discussion.

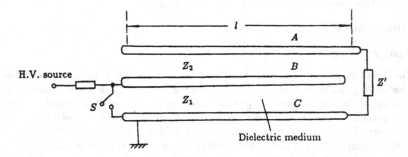

Fig.4-35 Basic structure of a Blumlein line in parallel-plates configuration, A=upper plate, B=middle plate, C=lower plate, S=H.V. switch, Z_1, Z_2 = characteristic impedances of the lower and upper lines, Z' = load impedance.

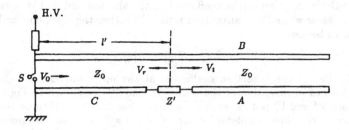

Fig.4-36 Equivalent circuit of the Blumlein line shown in Fig.4-35, with $Z_1 = Z_2 = Z_0$ where l' is the effective length of the line which includes part of the transmission (load) section.

In order to facilitate the discussion, let's re-draw the figure shown in Fig.4-35 into a different form in which we have set $Z_1 = Z_2 = Z_0$ (Fig.4-36). Before the switch is closed, the system may be viewed conceptually as two charged capacitors connected in parallel configuration via an impedance Z'. Let V_0 be the breakdown voltage of the switch, when the switch closes, a voltage pulse of amplitude V_0 is produced at the shorted end and propagates along the line toward the open end. If there was no discontinuity along the line, this voltage pulse would suffer no reflection until reaching the open end. However, the presence of the impedance Z' between plates A and C constitutes a discontinuity. When the voltage pulse reaches there reflection of a voltage pulse V_r occurs, meanwhile a voltage pulse V_t is transmitted to the open line on the right. The directions of propagation of these pulses are shown in Fig.4-36. To make the situation more visual, let us once again redraw the figure as that shown in Fig.4-37 in which we have fictitiously introduced some bending on the parallel plates. At $t = T_0$, the delay time of the line, the voltage pulse V_0 reaches the section rr', it sees a mismatched termination[18] of total impedance $Z' + Z_0$. As a result, a reflected pulse V_r of the same polarity but with an amplitude $\rho_v V_0$ is produced, where ρ_v is the voltage reflection coefficient at the section rr' and is equal to $\frac{Z' + Z_0 - Z_0}{Z' + Z_0 + Z_0} = \frac{Z'}{Z' + 2Z_0}$. From Section 4-2(a), we know that the total voltage across the section rr' at this time is

$$V = V_0 + \rho_v V_0 = 2V_0 \frac{Z' + Z_0}{Z' + 2Z_0} \qquad (4\text{-}71)$$

of which there is $V Z_0/(Z' + Z_0)$ across the characteristic impedance Z_0 of the open line on the right and $V Z'/(Z' + Z_0)$ across the load impedance Z'. Since the voltage across Z_0 does not suffer reflection at section rr', it transmits itself to the open line as a transmitted pulse V_t and propagates toward the open end. The voltage across the impedance Z' is absorbed in the load. At $t = 2T_0$, the reflected pulse V_r reaches the shorted end and suffers another reflection to become

$$V_r' = \rho_0 \rho_v V_0 = -\frac{V_0 Z'}{Z' + 2Z_0} \qquad (4\text{-}72)$$

where ρ_0 is the voltage reflection coefficient at the shorted end and is equal to -1 in this case. Meanwhile the transmitted pulse V_t reaches the open end and suffers a reflection to become

$$V_t' = \rho_l V_t = \frac{2Z_0 V_0}{Z' + 2Z_0} \qquad (4\text{-}73)$$

where ρ_l is the voltage reflection coefficient at the open end and is equal to 1. The directions of propagation of the pulses V_r' and V_t' are shown in Fig.4-37. At $t = 3T_0$, both V_r' and V_t' pulses reach the mismatched location. If these two pulses do not cancel each other completely, some reflection will occur at that point. Then the total voltage across the section rr' will have to be evaluated with the formula given in Eq.(4-45) for the cases of multiple reflection. One way to eliminate the multiple reflection is to make $V_r' = -V_t'$. Under such conditions, the total voltage across the section rr' can be expressed by the expression given in Eq.(4-71). The above derivations may be carried out with $Z_1 \neq Z_2$, the conclusion will be the same.

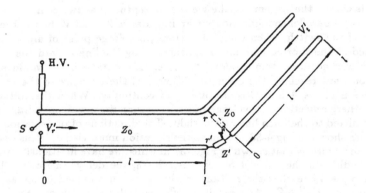

Fig.4-37 Same circuit of that shown in Fig.4-36 with imaginary bending on the plates.

Now let $Z' = 2Z_0$, from Eqs.(4-72) and (4-73) one can see that $V_r' + V_t' = 0$, i.e. the two pulses cancel each other at that point, then from (4-71) we obtain the total voltage across the rr' section as

$$V = \frac{3}{2} V_0 \qquad (4\text{-}74)$$

of which there is $(1/2)V_0$ across the impedance Z_0 and V_0 across the load Z'. Therefore, during the time interval $T_0 \leq t \leq 3T_0$, the voltage across the load Z' is a constant and equal to the input voltage V_0. For $t > 3T_0$, all succeeding reflections for this line continue to cancel in a similar manner, the voltage across the load Z' drops to zero as the capacitive energy initially stored in the line is now completely dissipated in the load (assuming Z' is resistive). As a result, the load sees a square voltage pulse of amplitude V_0 and duration $2T_0$ as shown in Fig.4-38(a). Under such condition, namely $Z_1 = Z_2 = Z_0$, and $Z' = Z_1 + Z_2$, the Blumlein line is said to be properly matched. If $Z' \neq Z_1 + Z_2$ or $Z_1 \neq Z_2$, from Eqs.(4-72) and (4-73), one can see that V_r' and V_t' will not cancel each other exactly, hence multiple reflection will occur. The voltage at the load will then appear as that shown in Fig.4-38(b) and Fig.4-38(c). Under such conditions, the line is said to be mismatched. To achieve maximum efficiency, the Blumlein line must be operated under the properly matched condition. Unless low efficiency is not the main concern, in the design of a Blumlein line one should try to make it match, i.e. $Z_1 = Z_2 = Z_0$ and $Z' = 2Z_0$. Otherwise, from Fig.4-38(a) and (b) one can see that a considerable amount of energy will not deposit into the load during the $T_0 \leq t \leq 3T_0$ period. As far as energy utilization is concerned, the amount of energy deposited into the load after $t = 3T_0$ in most cases is considered to be a loss and therefore is undesirable.

(b) Circuit Analysis

When the Blumlein line shown in Fig.4-35 is charged by a Marx generator, before the switch S is closed, the equivalent circuit of the system may be approximated by that shown in Fig.4-39. Since the value of R_l in general is much smaller than that of R_g, in a parallel connection as that shown in the figure, R_g can be neglected. Besides, the impedance Z' of a matched load is usually no more than a few ohms when water is employed as the line's dielectric, whereas the impedance of C_2 can be two orders of magnitude greater than the former, therefore Z' in the circuit can also be neglected. Now if we further substitute $C_1 + C_2 = C_l$, then the circuit shown in Fig.4-39 is reduced to a form identical to that shown in Fig.2-10. In Section 4-4(b), we have used that circuit to represent the system of a simple transmission line charged by a Marx generator and carried out an analysis for the system. Some analytic expressions of the concerned quantities were given in Eqs.(4-69) and (4-70), respectively. In this section we shall use these expressions to carry out a similar analysis for the system consisting of a Blumlein line charged by a Marx generator. The only modification that needs to be done is to replace $(C_1 + C_2)$ shown in Fig.4-39 by C_l. On this basis, the system shown in Fig.4-39 is analyzed under the following conditions.

(A) When $R = 0$ and $R_l = \infty$

Under the condition of $R = 0$ and $R_l = \infty$, from Eqs.(4-67) and (4-68), we have the three roots of Eq.(4-64)

$$\left.\begin{array}{l} \lambda_1 = 0 \\ \lambda_2 = j\omega \\ \lambda_3 = -j\omega \end{array}\right\} \tag{4-75}$$

where $\omega^2 = 1/(LC) + 1/(LC_l)$. Substituting Eq.(4-75) into Eqs.(4-69) and (4-70),

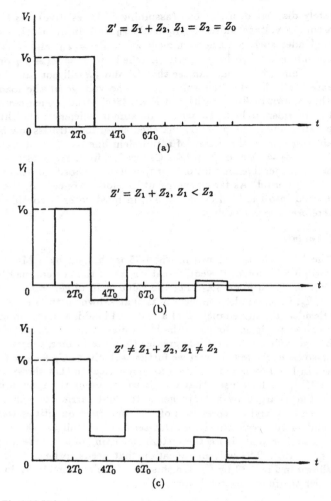

Fig.4-38 Voltage pulses appear at the load Z' under various conditions.

we obtain the voltage and current in the Blumlein line respectively as

$$V_l(t) = \frac{CV_m}{C + C_l}(1 - \cos \omega t) \tag{4-76}$$

and

$$i_2(t) = \frac{CC_l}{C + C_l}V_m\omega \sin \omega t \tag{4-77}$$

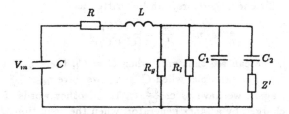

Fig.4-39 Approximate equivalent circuit of a Blumlein line charged by a Marx generator. V_m=peak voltage of the Marx generator, C =equivalent capacitance of the Marx generator, R =series resistance of the Marx generator, L =series inductance of the Marx generator, R_g=grounding resistance of the Marx generator, R_l=equivalent shunt resistance of the Blumlein line, C_1=capacitance between the center and lower plates (see Fig.4-35) of the Blumlein, C_2=capacitance between the center and upper plates, Z' =load impedance of the Blumlein.

The general behavior of V_l and i_2 as a function of time is shown in Fig.4-40. From the figure, we can see that when $V_l(t)$ is at its peak value, $i_2(t)$ is zero. Since the energy stored in the line is proportional to V_l^2 and the power is the product of V_l and i_2, that means at the moment the Blumlein line stores the maximum energy, the charging process stops. The times t at which the Blumlein line has maximum energy are $t = (2k + 1)\pi/\omega$ with $k = 0, 1, 2, \cdots$ etc. It appears that if the switch S shown in Fig.4-36 closes at these times, one may expect, the load Z' will receive the maximum energy from the line. This is so only under certain conditions. Let us see why it is so and what are the required conditions.

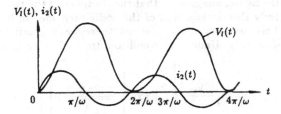

Fig.4-40 General behavior of V_l and i_2 as a function of time.

At $t = \pi/\omega$, the energy stored in the Blumlein line is

$$\epsilon_m = \frac{1}{2}C_l \left(\frac{2V_m C}{C + C_l} \right)^2 = \left(\frac{1}{2}CV_m^2 \right) \frac{4CC_l}{(C + C_l)^2} \tag{4-78}$$

where $(1/2)(CV_m^2)$ represents the maximum energy ϵ'_m of the Marx generator. The energy transfer efficiency, therefore, can be written as

$$\eta_\epsilon = \frac{\epsilon_m}{\epsilon'_m} = \frac{4CC_l}{(C + C_l)^2} \qquad (4\text{-}79)$$

From Eq.(4-79), one can see that only when $C = C_l$, then $\eta_\epsilon = 1$. Under other conditions, $\eta_\epsilon < 1$. For example, when $C \gg C_l$, we have $\eta_\epsilon \simeq \frac{4C_l}{C} \ll 1$, similarly, when $C \ll C_l$, again we have $\eta_\epsilon \simeq \frac{4C}{C_l} \ll 1$. In other words, for a system of Blumlein lines charged by a Marx generator, when the conditions $R = 0$, $R_l = \infty$ and $C = C_l$ are satisfied, the load Z' can receive the maximum energy from the energy source. This is, however, only for an ideal case. In practice, most situations are far from ideal. In the following, we shall discuss some of them.

(B) When $R > 0$ and $R_l = \infty$

Under the condition of $R > 0$ and $R_l = \infty$, the three roots of Eq.(4-64) become

$$\lambda_1 = 0$$
$$\lambda_2 = -\beta + \gamma$$
$$\lambda_3 = -\beta - \gamma$$

where $\beta = R/(2L)$, $\gamma^2 = \beta^2 - \omega^2$. From Eq.(4-66) we have

$$V_l(t) = \frac{CV_m}{C + C_l} \left\{ 1 - \frac{1}{2} e^{-\beta t} \left[(e^{\gamma t} + e^{-\gamma t}) + \frac{\beta}{\gamma}(e^{\gamma t} - e^{-\gamma t}) \right] \right\} \qquad (4\text{-}80)$$

The time-dependent behavior of V_l is determined by the property of γ. If γ is real, i.e. $\beta \geq \omega$, $V_l(t)$ is non-oscillatory. When γ is imaginary, i.e. $\beta < \omega$, V_l is oscillatory. The general behaviors of the three possible modes are shown in Fig.4-41. From the figure, one can see that the risetimes of the two non-oscillatory results are relatively slower than that of the oscillatory one. In high power pulse applications, fast rise time is generally desired, therefore, we shall discuss the latter case, namely $\beta < \omega$, only. Under this condition, the real value of V_l is

$$V_l = \frac{CV_m}{C + C_l} \left\{ 1 + \exp\left[\frac{-\pi}{\sqrt{\left(\frac{\omega}{\beta}\right)^2 - 1}} \right] \right\}$$

hence the energy transfer efficiency becomes

$$\eta_\epsilon = \frac{CC_l}{(C + C_l)^2} \left\{ 1 + \exp\left[\frac{-\pi}{\sqrt{\left(\frac{\omega}{\beta}\right)^2 - 1}} \right] \right\}^2 \qquad (4\text{-}81)$$

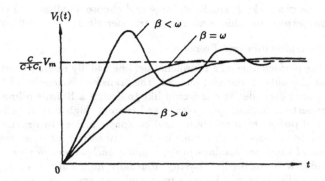

Fig.4-41 General behavior of $V_l(t)$ under three different conditions.

If we make $R \to 0$, then the right hand side of Eq.(4-81) becomes identical to that of Eq.(4-79) which has higher efficiency. This example once again illustrates that one way to obtain higher energy transfer efficiency is to minimize the series resistance of the Marx generator.

(C) $R = 0$, $R_l \neq \infty$, and $R \neq 0$, $R_l \neq \infty$

For the cases of $R = 0$, $R_l \neq \infty$ and $R \neq 0$, $R_l \neq \infty$, their respective energy transfer efficiency can be expressed as

$$\eta_\epsilon \simeq \frac{4CC_l}{(C + C_l)^2} \left[1 - \frac{3\pi C}{\omega R_l (C + C_l)^2}\right] \tag{4-82}$$

and

$$\eta_\epsilon' \simeq \frac{CC_l}{(C + C_l)^2} \left[1 - \frac{3\pi C}{\omega R_l (C + C_l)^2}\right] \left\{1 + \exp\left(\frac{-\pi}{\sqrt{\left(\frac{\omega}{\beta}\right)^2 - 1}}\right)\right\}^2 \tag{4-83}$$

If we let $R \to 0$ and $R_l \to \infty$, both η_ϵ and η_ϵ' once again become identical to that of Eq.(4-79).

The above illustrations indicate that when $R \to 0$ and $R_l \to 0$, the energy transfer efficiency between the Marx generator and the Blumlein line will become higher. If we further make $C = C_l$, then the efficiency becomes 100%. That, of course, is only ideal. In practice neither $R = 0$ nor $R_l = \infty$ can be truly realized. Even the condition $C = C_l$ is quite difficult to achieve. Therefore, in real cases, the energy transfer efficiency that can be achieved is always considerably less than ideal. How much less depends on the relative magnitudes of the various parameters involved. If one can make R small, R_l large and choose a value of C_l close to that of C, one may expect to achieve an energy transfer efficiency of 80% or so.

(c) Design Consideration and Procedure

To rate the performance of a Blumlein line, one usually refers to certain parameters such as the output voltage, current, the transfer efficiencies of the voltage, energy and power etc. Ideally one would like to design a Blumlein line having the following properties: high output voltage and current, high transfer efficiencies for the energy and power, high reliability and compact in size. In practice, however, it is not possible to design such a machine that can satisfy all these requirements, because some of them are fundamentally conflict each other. For example, if one wishes to build a Blumlein line having unusually high output voltage, then its energy transfer efficiency will become poor and vice versa. So, in most practical cases, one either follows the balanced approach in which one takes every aspect into consideration and aims at fair but not excellent results for most of them, or the extreme approach in which one aims at excellency for a certain particular aspect and ignores most of the rest. In the following we shall use examples to illustrate each of these approaches.

This example is to illustrate the application of the balanced approach to design a Blumlein line having the following specifications:

Output Voltage	1 MV
Load Impedance	6 Ω
Pulse Duration	85 ns

In order to get fair results for most aspects, it is essential to have a matched line. The following discussion is based on a properly matched Blumlein line. From the specifications given above, we know that the characteristic impedance of the line in this case is 3 Ω. This suggests water to be the most appropriate insulator for such low impedance [see Eq.(4-49) and Table 3-2]. Once ϵ_r is given, we can from Eq.(4-41) determine the effective length l' [see Fig.(4-36)], namely

$$l' = \frac{cT_0}{\sqrt{\epsilon_r}} \tag{4-84}$$

where c is the speed of light, ϵ_r the relative permittivity of water, and T_0 the delay time of the line and equal to 85/2 ns in this case. The effective length l' of the line differs from l by an amount of Δl, the half length of the transmission (load) section of the line, i.e. $l' = l + \Delta l$. In the previous discussions, we have assumed an ideal situation and taken Δl to be equal to zero. In the present case, Δl is to be determined from the values of r_1, r_2 and r_3. These r values in turn can be determined by Eq.(4-49) in conjunction with Eq.(3-18) for E_b, the breakdown field under uniform conditions in liquid and the relation

$$E_i = \frac{V_m}{r_i \ln(r_2/r_1)} \tag{4-85}$$

where $i = 1, 2$, V_m is the maximum charged voltage to the line, and E_i is the field strength at $r = r_i$. The parameter t appears in Eq.(3-18) may be expressed by[19]

$$t = \frac{\pi}{2} \left[\frac{LC(C_2 + C_1)}{C + C_1 + C_2} \right]^{1/2} \tag{4-86}$$

In Eq.(4-86), L and C are respectively the inductance and capacitance of the Marx generator and in this case $L = 16 \times 10^{-6}$H, $C = 20 \times 10^{-9}$F. C_1 and C_2 are the respective capacitance of the inner and outer lines of the Blumlein line and can be determined by the relation

$$C_i = \frac{l_i}{Z_0} \frac{\sqrt{\epsilon_r}}{c} \tag{4-87}$$

where $i = 1, 2$ and c is the speed of light, Z_0 is the characteristic impedance of the Blumlein line and in this case $Z_0 = (Z'/2) = 3\Omega$, l_i is the length of line i [see Eqs.(4-30) and (4-41)]. In order to assure reliability and prevent misfire, one usually sets some safety factor in the design and let

$$E_i = \xi E_b \tag{4-88}$$

So that the actual field strength E_i is always less than the breakdown field strength E_b if ξ is chosen to be less than unity. In summing up, we have six independent relations [Eqs.(4-84), (4-85), (4-87) and (4-88)] which involve six unknown parameters (l', l_1, l_2, r_1, r_2, r_3). Thus the six unknown parameters can be determined. Further by using the relationship[19]

$$R_l = \epsilon_0 \epsilon_r \eta / c$$

one can obtain the value of the bleeding resistance of the Blumlein line, where η is the resistivity of pure water.

At this point, all the essential parameters for the construction of the specified Blumlein line are obtained. By using these parameters in conjunction with other given parameters, one can readily obtain other pertinent relations such as the energy transfer efficiency etc. The results so obtained are summarized in Table 4-3 to 4-5. A schematic diagram of the Blumlein line so designed by Zhao et al[19] is shown in Fig.4-42. The results given in Table 4-3, indicate that the Blumlein line basically satisfies the conditions as a properly matched line. From the previous discussion we know that a matched line can yield maximum energy transfer efficiency. If we substitute the values of Z', Z_1 and Z_2 given in Table 4-3 into the relation[19]

$$\eta_{BL} = \frac{16 Z' Z_1 Z_2}{(Z' + Z_1 + Z_2)^2 (Z_1 + Z_2)} \tag{4-89}$$

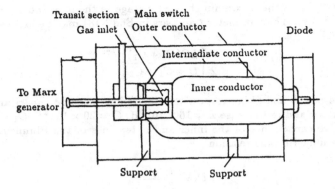

Fig.4-42 Schematic diagram of the Blumlein line designed by Zhao et al[19] with the parameters shown in Table 4-3.

Table 4-3 Calculated parameters for the design of the specified Blumlein line.

	r_{out}/r_{in} (cm)	Impedance $Z_i(\Omega)$	Length l_i(cm)	Capacitance C_i(nF)	Bleeding resistance $R_l(k\Omega)$
Inner Line	43.8/28	3	90.3	8.988	1.2
Outer Line	69/44.6	2.93	111.3	11.359	0.93
Transmission Load Section	69/28	5.93	39.7	1.96	5.42

Table 4-4 The ratio of the real electric strength to the allowable electric field at each cylinder surface

	E(kV/cm)	$\xi = E/E_{allowable}$
Outer-surface on inner cylinder	79.8	0.565
Inner-surface on intermediate cylinder	51.0	0.198
Outer-surface on intermediate cylinder	51.4	0.2
Inner-surface on outer cylinder	33.2	0.27

we will find that the energy transfer efficiency between the Blumlein line and the load is nearly $\eta_{BL} = 1$, indicating nearly all the energy in the line is transferred to the load. Should the line have been mismatched, such a result would not be possible. A further examination of the results presented in Table 4-3 to 4-5, one can readily see that the general behavior of the Blumlein line in most aspects are fairly good. This is largely the consequence of the so-called balanced approach that has been followed in the design. In the following, we shall discuss another approach in which the design is based on the principle of achieving excellency in a certain particular area and ignoring most of the rest.

Table 4-5 Measured parameters of the Blumlein line mentioned in Table 4-3.

Charging voltage	Blumlein output voltage		Diode peak current	
V_m(kV)	V_i(kV)	τ(ns)	I_0(kA)	τ(ns)
1260	904	85	119	100

In certain applications, high output voltage is required, e.g. in the production of high energy electron beam or intense X-ray. In such cases, one is required to design a Blumlein line with $Z' > (Z_1 + Z_2)$, as from Fig.4-38(c) one can see that under such conditions, the output voltage can be significantly higher than the input one. To do so, however, some price has to be paid, i.e. one only can have a mismatched line of relatively low energy efficiency. Let us assume that the Blumlein line to be designed is required to have the following specifications:

Input Voltage	10 MV
Output Voltage	13 MV
Characteristic impedances	$Z_1 = 12\Omega, Z_2 = 22\Omega$
Cylindrical configuration	

Let's begin the procedure with Eq.(4-85) and assume that the relation

$$E_3 = \frac{V_m}{r_2 \ln\left(\frac{r_3}{r_2}\right)}$$

is still valid (strictly speaking, it is not) and it is related to E_2 by

$$E_3 = \delta E_2 = \frac{\delta V_m}{r_2 \ln(r_2/r_1)} \tag{4-90}$$

where $\delta < 1$. Further let $r_2/r_1 = x$, then from Eq.(4-90) we have

$$r_3/r_2 = x^{1/\delta} \tag{4-91}$$

thus

$$Z_1 = 60 \ln x/(\sqrt{\epsilon_r}) \quad \text{and} \quad Z_2 = 60 \ln x/(\delta\sqrt{\epsilon_r}) \tag{4-92}$$

Since the voltage and impedances involved are high, it is appropriate to use oil rather than water as the insulator, hence $\epsilon_r = 2.2$ in this case. Substituting the values of Z_1, Z_2 and ϵ_r into Eqs.(4-92), we obtain

$$x = 1.345 \quad \text{and} \quad \delta = 0.545 \tag{4-93}$$

If we let $r_1 = 111$ cm, then from Eqs.(4-93) and (4-91), we get $r_2 = 149.3$ cm and $r_3 = 257.2$ cm. During the time $T_0 \leq t \leq 3T_0$, the voltage across the load Z' is

$$V_l = V_m(1 + \rho_v)\frac{Z'}{Z' + Z_2} = V_m \frac{2Z'}{Z' + Z_1 + Z_2}$$

hence $V_l/V_m = 1.3 = 2Z'/(Z'+34)$ which yields $Z' = 63.1\Omega$. At this point, we have obtained all the essential parameters, except the length l, for the construction of the Blumlein line specified at the beginning. The length l can be easily determined via Eq.(4-84), once the pulse duration is chosen. Alternatively, one may use Eqs.(3-18) and (4-88) to determine l. From this example, we can see that for a high impedance line, oil is more appropriate than water to be used as the insulator. However, if even higher output impedance, say several hundred Ω, is required, then one will have to rely on other techniques such as using a Blumlein line with a helical conductor[20] to achieve the goal.

4-6 Stacked Lines and Magnetically Insulated Lines

(a) Stacked Transmission Lines

When several transmission lines are connected together, the system is called a stacked line. Both the simple transmission line and Blumlein line discussed in the preceding sections can be stacked to yield either high voltage or current depending on the way they are structured. However, in most pulsed power applications, stacked transmission lines have been often used to enhance the voltage rather than the current. Therefore, in this section we shall mainly discuss the aspects concerning the stacked line as a voltage multiplier. There are various ways to design such a voltage multiplier. The sketch shown in Fig.4-43 is one typical example of it[21]. In this example, three identical coaxial lines are used for convenience. In principle, the number of lines and the type of lines may be used in a stacked line are unlimited. To facilitate the discussion, we replace the sketch shown in Fig.4-43(a) by an equivalent circuit as shown in Fig.4-43(b). Assuming at $t = 0$ the switch S closes, then a voltage pulse of amplitude V_0 is initiated at the open end A in each line. These pulses propagate along the lines toward the end B. When at $t = T_0$ they arrive there, they all suffer a reflection. Because lines 1, 3, 5 have open ends at B whereas lines 2 and 4 have closed ones, the total voltages at the end B become $2V_0$ for the former and zero for the latter as indicated in the figure. These pulses will last until $t = 3T_0$ at which time reflected pulses initiated from other ends will arrive there and they will be distorted unless the reflected pulses cancel each other exactly. Since these voltages are in series configuration, the current i going through Z' can be expressed as

$$i = \frac{6V_0}{3Z_0 + Z'}$$

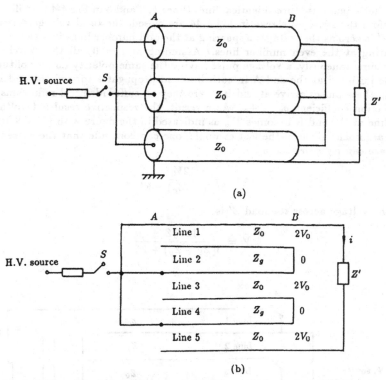

(a)

(b)

Fig.4-43 Three identical simple transmission lines stacked together with Z' as the load. (a) the actual arrangement. (b) the equivalent circuit of (a).

and hence the voltage across the load Z' can be written as

$$V_l = iZ' = \frac{6V_0 Z'}{3Z_0 + Z'}$$

If the number of lines employed is n, then $V_l = 2nV_0 Z'/(nZ_0 + Z')$ and for a matched load $Z' = nZ_0$, then

$$V_{lm} = nV_0 \qquad (4\text{-}94)$$

That means, under ideal conditions, one can expect a voltage gain of n from such a stacked line. In practice, however, the real gain one can achieve in general is considerably less than that, due to various practical reasons.

As can be seen from Fig.4-43, in the last example the entire system is controlled by a single switch S. If each individual line is switched on by a separate switch as indicated in Fig.4-44, the voltage gain obtained from such a multiple switching system can be twice as high as that from the single switching system. To see how it works, let's again use three identical lines (lines 1, 3 and 5 in Fig.4-44) to illustrate it. Before the switches shown in Fig.4-44 are closed, the total voltage across the load Z' is zero as the voltages appearing at the odd number lines just cancel those appearing at the even number lines. Assuming at $t = 0$, all the switches are closed simultaneously, a voltage pulse having the same polarity and amplitude V_0 will be initiated at the end A in each line and propagates toward the end B. At $t = T_0$, these pulses arrive at end B where they all suffer identical reflections with a reflection coefficient $\rho_v = +1$. As a result, the respective resultant voltage of each line at the end B becomes $2V_0$ as indicated in the figure with ($\uparrow\uparrow$). With the same argument given in the last example, one can conclude that the current i in this case can be expressed as

$$i = \frac{12V_0}{3(Z_0 + Z_g) + Z'}$$

and the voltage across the load Z' is

$$V_l = \frac{12V_0 Z'}{3(Z_0 + Z_g) + Z'}$$

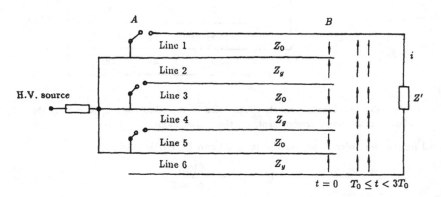

Fig.4-44 Three identical simple transmission lines stacked together and each individual line is controlled by a separate switch, where Z' is the load.

If the number of lines used is n and $Z_g = Z_0$, then

$$V_l = \frac{4nV_0 Z'}{2nZ_0 + Z'}$$

For a matched load $Z' = 2nZ_0$, V_l becomes

$$V_{lm} = 2nV_0$$

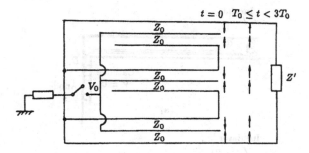

Fig.4-45 A stacked system of three Blumlein lines controlled by a single switch.

In a comparison with Eq.(4-94), one can see that, if other conditions are identical, a multiple switching system can have voltage gain twice of that for a single switching system.

In the above example, the stacked systems are assembled with simple transmission lines. One can equally use Blumlein lines as the basic unit to form stacked lines. The sketch shown in Fig.4-45 is a stacked line formed from three Blumlein lines and controlled by one switch. In addition, there are a number of other forms of stacked lines which have been employed in pulsed power applications. Each of these lines has its own particular feature and can serve special purpose. The notable ones are: the multistage wave generator (MSWG)[22], the cumulative wave line and darlington network[23], the parallel stacked Blumlein line[24], the Archimedes' voltage multiplier[13], the split-diamond convolute structure[25], the polarity converter[26], the tri-plate water stripline[27] and the repetitive stacked transmission lines[28,29]. Theoretical analysis of these stacked lines may be carried out by means of similar procedures described in the two previous examples. For detailed information concerning these lines, one may refer to the respective references given at the end of this chapter.

(b) Magnetically Insulated Transmission Line

In comparison with the basic principle of the various transmission lines discussed in the preceding sections, magnetically insulated transmission line (MITL) is structured and operated on an entirely different idea. Its basic principle is, by utilizing the self magnetic field, to inhibit the electrons in a vacuum transmission line to reach the anode such that high insulation between the two conductors can be maintained. Because of such property, MITL can be built in a very compact form, yet it is still capable of transmitting a large amount of power to a small target at a far distance. Theoretical study of the subject has been carried out by many investigators. The basic approach may be summarized into two main categories: the relativistic treatments[30] and the relativistic Brillouin Version[31]. In the present discussion, we shall use the results of the latter approach only as

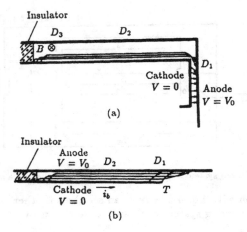

Fig.4-46 (a) magnetically load-limited flow; (b) magnetically self-limited flow [32].

they are expressed in terms of analytic forms. According to the theory, there are three types of magnetic cutoff of electron flow, of which two of them are applicable to the pulse charged transmission lines. They are the load limited and self limited magnetic cutoffs and both are illustrated in Fig.4-46. In the former type, the electrons flow along equi-potential surfaces and the current flowing to the anode in region D_1 provides the magnetic field which cuts off electrons flow to the anode in region D_2, whereas in the latter type the magnetic cutoff in region D_2 is caused by current flowing to the anode in region D_1 and the total current. The boundary between D_1 and D_2 is determined by the magnetic cutoff process itself. The above two types of magnetic cutoffs can also be illustrated with the flow pattern shown in Fig.3-26 in which the electron trajectories are roughly semi-circles and return to the cathode before touching the anode in region D_1. The nature of the electron trajectories is determined by the geometric configuration of the transmission line as well as by the applied voltage and current. There is a critical condition at which the trajectories of the electrons are tangent to the anode, i.e. the electrons are inhibited to reach the anode. In quantitative terms, the critical condition may be stated as follows. When the total current I in the transmission line is greater than the minimum current I_l, electron flow to the anode will be magnetically cutoff, where I_l is expressed as[32]

$$I_l = I_\alpha g \gamma_l^3 \ln[\gamma_l + (\gamma_l^2 - 1)^{1/2}] \qquad (4\text{-}95)$$

In the expression, $I_\alpha = 8500$ A and γ_l is related to the anode potential V_0 by

$$\gamma_l + (\gamma_l^2 - 1)^{3/2} \ln[\gamma_l + (\gamma_l^2 - 1)^{1/2}] = \frac{eV_0}{m_0 c^2} + 1 \qquad (4\text{-}96)$$

in which e and m_0 are respectively the electronic charge and rest mass, c is the speed of light, g is a geometrical factor and its analytic forms for the various configurations shown in Fig.4-47 are given in Table 4-6.

Table 4-6 Geometrical factors g [32].

Electrode configuration	Geometrical factor g
Parallel planes [Fig.4-47(a)]	$y_0/(2\pi x_0)$
Concentric cylinders [Fig.4-47(b)]	$[\ln(R_2/R_1)]^{-1}$
Concentric cones [Fig.4-47(c)]	$[\ln(\tan(\theta_0/2)) - \ln(\tan(\delta/2))]^{-1}$

An examination of Eqs.(4-95), (4-96) and the analytic forms of g given in Table 4-6, one can see that I_l may be expressed in terms of the anode potential V_0 and the geometrical dimensions of the transmission line. In the case of a cylindrical transmission line, we have

$$I_l = F(V_0, R_2, R_1) \qquad (4\text{-}97)$$

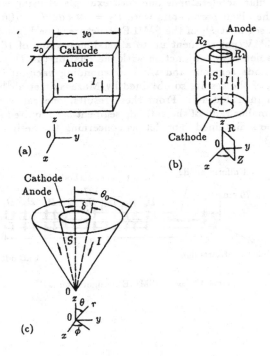

Fig.4-47 Various geometrical configurations employed by Creedon for the study of magnetically cutoff flow [32].

Table 4-7 SMILE parameters (Reproduced from Ref.34)

MITL Segment i	Segment voltage V_i(MV)	Optimum cathode r_i(cm)	Vacuum impedance $Z_i(\Omega)$	Operating impedance $Z_i(\Omega)$
1	2.0	13.33	45.3	30
2	4.0	8.05	55.0	40
3	6.0	6.30	80.0	61
4	8.0	3.86	96.0	80
5	10.0	2.69	121.0	97
6	12.0	1.90	149.0	122
7	14.0	1.35	162.0	135
8	16.0	0.952	180.0	151

When any three of the four parameters I_l, V_0, R_2, R_1 are given, from Eq.(4-97) one can determine the fourth one. On this basis, large size MITL has been designed and successfully operated in recent years[33]. Design of the SMILE structure for the modified Radlac accelerator is one such example of using this procedure[34]. In the design, the given parameters were $R_2 = 19$ cm, $I = 106$ kA and $V_0 = 2$ MV for the first segment D_1 of the SMILE structure. (see Fig.4-48). V_0 increases successively at 2MV/per segment up to a maximum value of 16 MV at D_8. By using these parameters in conjunction with the relation (4-97), the values for the radii of the cathode segment and the corresponding vacuum impedances were obtained. Some of the results so obtained by Mazarakis et al[34] are reproduced here and shown in Table 4-7. From these results, one can see that at higher anode voltage, smaller radii of the cathode segment are required so that magnetic insulation can be maintained. For details concerning the design work, one could to refer to Ref.[34].

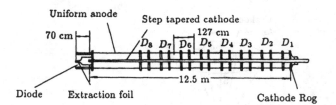

Fig.4-48 Radlac/SMILE configuration [34].

References of Chapter 4

[1] F. E. Terman and J. M. Pettit, *Electronic Measurements*, (2nd ed. McGraw-Hill, New York, 1952).

[2] Adolf J. Schwab, *Field Theory Concepts*, (Springer-Verlag, World Publishing Corp., Berlin, 1988).

[3] R. E. Matick, *Transmission Lines for Digital and Communication Networks*, (MeGraw-Hill, New York, 1969).

[4] J. D. Kraus, *Electromagnetics*, (McGraw-Hill, New York, 1953).

[5] K. G. Black and T.J. Higgins, *IRE Trans, Microwave Theory Tech.* **MTT-3**, (1955), 93.

[6] F. G. Primozich, *Millimicrosecond Pulse Studies-Engineering Design of Tapered Transmission Lines*, PhD. Thesis, Carnegie Institute of Tech. Pittsburgh Pa., 1954.

[7] J. A. Nation, *Particle Accelerators*, **10**, (1979), 1.

[8] J. Benford et al, *IEEE Trans, on Nuclear Sci.*, NS-26, (1979), 4208.

[9] J. J. Ramirez, *14th IEEE Power Modulator Symp.*, (1980), 314.

[10] Ion Physics Corp., Burlington, Massachusetts, Report on Development of five EMP generators, Contract #29601-69-c-0138 (1970).

[11] Ian Smith, *7th IEEE Pulsed Power Conf.*, (1989), 103.

[12] Qi Zhang et al., *5th Int. Symp. on High Voltage Engineering*, (1987), p.91.04.

[13] Wang Ying, *Electrical Source for High Power Pulse Applications* (in Chinese), (Chinese Atomic Energy Press, Beijing, 1991).

[14] N. Camarcat et al., *Laser and Particle Beams*, **3**, No.4, (1985), 415.

[15] B. Z. Hollmann, *6th IEEE Pulsed Power Conf.*, (1987), 664.

[16] W. N. Weseloh, *7th IEEE Pulsed Power Conf.*, (1989), 989.

[17] P. Corcoran et al., *7th IEEE Pulsed Power Conf.*, (1989), 465.

[18] J. H. Crouch, W. S. Risk, *Rev. Sci. Instrument*, **43**, (1972), 632.

[19] Zhao Zhonghong et al., *J of Electronics*, **6**, (1989), 162.

[20] T. Teranishi et al., *8th IEEE Pulsed Power Conf.*, (1991), 315.

[21] I. A. D. Lewis, *Electr. Eng.*, **27**, (1955), 332.

[22] J. L. Harrison, *4th IEEE Pulsed Power Conf.*, (1983), 277.

[23] I. D. Smith, *IEEE 15th Power Modulator Symp.*, (1982), 223.

[24] S. Shope et al. DASA, (1970), 2482.

[25] R. B. Miller, *IEEE Tran, Nuclear Sci.*, NS-32, (1985), 3149.

[26] M. T. Buttram, SAND 83-0957 1983.

[27] T. H. Martin et al., *3rd Int. Topical Conf. on High Power Electron and Ion Beams*, Novosibirsk, (1979).

[28] F. Davanloo et al., *8th IEEE Pulsed Power Conf.*, (1991), 971.

[29] C. A. Pirrie et al. *8th IEEE Pulsed Power Conf.*, (1991), 310.

[30] A. Ron et al. *IEEE Trans. Plasma Sci.*, PS-1, (1973), 85.

[31] J. M. Creedon, *J. Appl. Phys.*, **46**, (1975), 2946.

[32] J. M. Creedon, *J. Appl. Phys.*, **48**, (1977), 1070.

[33] R. C. Pate et al. *6th IEEE Pulsed Power Conf.*, (1987), 478.

[34] M. G. Mazarakis et al. *8th IEEE Pulsed Power Conf.*, (1991), 86.

CHAPTER 5
SWITCH

5-1 Switch Fundamentals

In high power pulse applications, switches capable of handling tera-watt (10^{12} W) power and having jitter time in the nanosecond (10^{-9} s) range are frequently needed. The conventional switches, such as the high voltage switch gears are no longer adequate to meet these requirements. Development of new types of switch thus become necessary. On the basis of the technique employed for transferring the energy, high power switches may be classified into two basic types: closing and opening switches. Typical applications of these switches are shown in Fig.5-1. When closing switch is employed, the electrostatic energy is usually accumulated from the primary energy source at relatively low power level and stored in a capacitor C while the switch S_c is kept open as indicated in Fig.5-1(a). The stored

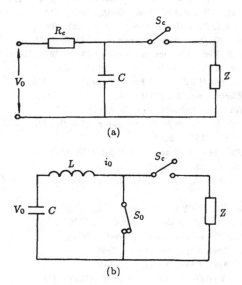

(a)

(b)

Fig.5-1 Typical circuit for the application of (a) closing switch and (b)opening switch V_0– high voltage source, R_e–charging resistor of high resistance, C–capacitor, Z– load, L–inductor, S_c–closing switch, S_0–opening switch.

energy is proportional to the square of the voltage V of the capacitor and is $\epsilon_0 = (1/2)CV^2$. When the switch S_c closes, a certain amount of the stored energy will be transferred to the load Z. The energy transfer efficiency η_ϵ is determined by many factors which include the characteristics of the load, the circuit parameters and the properties of the switch S_c. If the load is capacitive, η_ϵ can be estimated from Eq.(4-83). The maximum energy transfer efficiency in this case can be 100%. Of course, that is only in theory. In practice, 80% or so probably is the maximum efficiency achievable. In the case of the opening switch is used, the electrostatic energy stored in the capacitor C is first transformed into magnetic form and stored temporarily in an inductor L while switch S_0 is kept closed and switch S_c is opened as shown in Fig.5-1(b). At a proper time, i.e. when the current i_0 reaches a desired value, the initially closed switch S_0 is opened and the switch S_c closes, some of the magnetic energy stored in L is then transferred to the load Z. The total energy stored in L at this time is $\epsilon_0' = (1/2)Li_0^2$. The energy transfer efficiency is again dependent on many factors. If the load is an inductor of magnitude L_l, under an ideal situation the energy transfer efficiency can be expressed as

$$\eta_\epsilon' = \frac{LL_l}{(L + L_l)^2} \tag{5-1}$$

From Eq.(5-1), one can see that when $L = L_l$, the efficiency is maximum and equal to 25%. As compared to the efficiency of the closing switch circuit shown in Fig.5-1(a), this is clearly a rather poor approach as far as transferring energy is concerned. The main advantage of the opening switch circuit (inductive system) is that it can achieve significantly higher energy density than that of the capacitive system (for more details about opening switches, one may refer to Section 2-4). Generally speaking, closing switches have been used much more widely in high power pulse applications than the opening switch.

The actions of closing and opening the switches can be initiated either by external triggering or by self-break process. The basic principle of switching is relatively simple: at a proper time, change the property of the switch medium from that of an insulator to that of a conductor or the reverse. To achieve this effectively and precisely, however, is rather a complex and difficult task. It involves not only the parameters of the switch and circuit but also many physical and chemical processes. Although most of the phenomena associated with switching have been extensively studied, there are many areas which are still not well understood. To supplement the lack of sufficient knowledge in certain areas, designers frequently have to rely on empirical data and scaling laws to help their design. There is no unique formula or recipe that can be universally employed to design all types of switch. Designs of switch are largely carried out on the individual basis, i.e. most switches are designed to meet certain specific purpose and operate on different principles. For this reason, development of switch technology has been rather diverse and many types of switches with different characteristics have been developed in the past. To meet the various requirements many new ones are still under investigation. As pointed out before, proper design of an optimum switch requires knowledge in many areas. By and large, the property of the medium employed between the switch electrodes probably is the most important factor that determines the performance of the switch (except for vacuum switches). As gas, liquid and solid all can be used as the switch medium, switches are frequently classified as gas switch (includes plasma switch), liquid switch and solid switch (includes magnetic switches). Switches also can be classified on the basis of other parameters and properties such as: self-breakdown and externally triggered switches on the basis of trigger mechanism; statically charged and pulse charged switches on the basis of charging mode; single channel and multi-channel switches on the basis of the number of conducting channels; volume discharge and surface discharge switches on the basis of the discharge property and so on. As high power switches are so much diverse in their characteristics, there is no unique way to describe them by using one set of parameters. The only parameters that are common to most switches are illustrated in Fig.5-2. Each letter in the figure represents one of such parameter and the meanings of these parameters are described below.

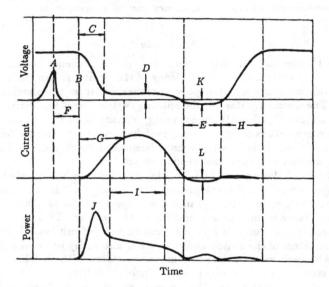

Fig.5-2 Characteristics of voltage, current and power for a steadily charged, externally triggered switch. A=trigger pulse, B=hold off voltage, C=Voltage fall time, D=conduction drop, E=recover time, F=delay time, G=current rise time, H=recharge time, I=current pulse width, J=peak power, K=reverse voltage, L=deionization current.

A. Trigger pulse —a fast pulse supplied externally to initiate the action of switching, the nature of which may be voltage, laser beam or charged particle beam.

B. Hold-off voltage—the maximum static voltage V_s that can be applied to the switch before breakdown between the main electrodes occurs. If the switch is pulse charged, the hold-off voltage can be greater than V_s and its magnitude depends on the risetime of the applied voltage pulse.

C. Voltage fall time—after breakdown is initiated, the time interval during which the voltage drops from the value of hold-off to that of the conduction drop. For gas switches, this roughly corresponds to the resistive phase during closure.

D. Conduction drop—the voltage drop across the switch impedance during conduction.

E. Recover time—the time interval during which the voltage reverses its polarity.

F. Delay time—the time interval between the time the trigger pulse is at its peak value and the point at which switch starts to close or open.

G. Current rise time—the time interval required for the current to rise from the 10% to 90% of its peak value.

H. Recharge time—the time interval between the end of the recover time and the point at which the voltage recovers to the hold-off value.

I. Current pulse width—the time duration corresponds to the full width half maximum of the current pulse.

J. Peak power—the maximum value of the product of voltage and current which occur at the same time (alternatively it is also defined by some authors as the product of the peak current and the hold-off voltage which are not necessarily occurring at the same time).

K. Peak reverse voltage—the maximum value of the reversed voltage.

L. Peak reversed current—the maximum value of the reversed current.

M. Energy transferred—the time integral of the product of voltage and current.

N. Life time—under normal operating conditions, the total number of switching operations beyond which the switch can no longer function properly.

O. Total charge transferred—the accumulated total charge that has passed through the switch during its life time.

P. Reliability—a measure of the ratio between the number of successful operations to the total number of operations within the life time of the switch.

As mentioned before, the basic principle of switching is to change the property of the switch medium from that of an insulator (high resistivity) to that of a conductor (low resistance) or the reverse. Accompanying the process there is usually transformation of state of the medium taking place. For instance, during the switching process of a solid gap, the medium initially is in solid state and subsequently transforms into liquid, gas and finally into the state of plasma. Meanwhile the resistance of the switch decreases accordingly and usually by several orders of magnitude. If a gas medium is employed such as in a gas spark gap, the state transformation of the medium would involve only from gas to plasma. In these two instances, both of the final states, which conduct most of the current are plasmas. However, the currents per unit area (current density) they conducted are significantly different. From Fig.5-3, one can see that the solid gaps in general can conduct far more current than the gas gaps. This is because both the current density and energy density are dependent not only on the property of the medium, but also on the geometrical dimensions of the plasma. This point can be further illustrated with the data presented in Fig.5-3 for the gas gaps and volume discharge in gas. In these two instances, the basic switching processes are similar. The difference is that, in the gas gaps, current is mainly conducted by filamentary plasma channels whereas in the volume discharge it is conducted by volumetric plasma of large dimensions, hence the current density is low. For this reason, it is always desirable, without sacrificing other requirements, to keep the design of the switch compact. By so doing, one can also minimize the switch inductance which has great influence on the rise time of the current pulse.

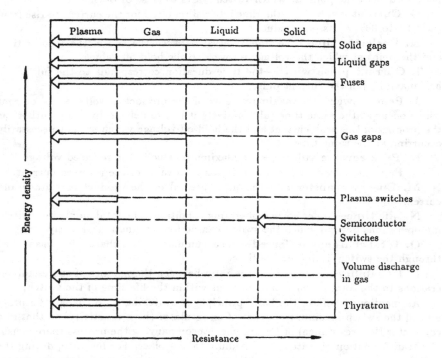

Fig.5-3 Schematic chart showing the ranks of various switches in the energy density scale and state transformation.

Fig.5-3 is a chart showing the ranks of the various types of switches in the energy density scale. It also illustrates how the switch medium changing its state during switching. Certain types of switch operate in one state only, e.g. plasma switches, semiconductor switches and thyratrons. From the chart one can see that the one-state switches fall into the low ranks in the energy density scale whereas all the three types of high rank switches, namely, solid gaps, liquid gaps and fuses involve successive transformations of state during switching. This indicates that state transformation of the switch medium is one of the basic elements in achieving high energy density switching. How to achieve it effectively has been one of the major problems in the development of switch technology. It requires both proper understanding of the physical and chemical processes involved as well as optimum design of the switch and the circuit employed. Detailed discussion of these aspects is deferred to some later sections, first let us, in the general sense, examine some other properties of the various types of switches.

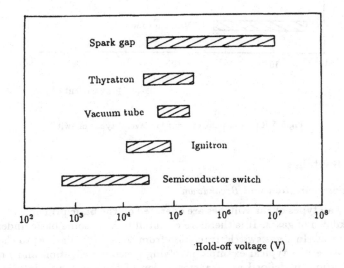

Fig.5-4 Range of hold-off voltage (including pulse charged mode) for several types of switch.

Fig.5-4 shows five types of switches in a chart of hold-off voltage range. Fig. 5-5 is a similar chart but showing the peak current range for the same group of switches. From these charts, one can see that, as far as hold-off voltage and peak current ranges are concerned, spark gaps are by far the most attractive ones. In the energy density scale, they also rank high as can be seen in Fig.5-3. For these reasons, we shall in the following sections devote considerably more time to discuss spark gap switches. However, these discussions shall remain in condensed form and deal with only the essential elements. For more comprehensive discussions on this subject, one can look up Refs.[1,2].

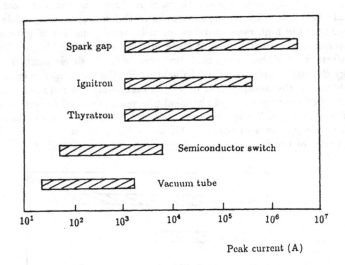

Fig.5-5 Range of peak current for several types of switch.

5-2 Gas Switches

(a) Gaseous Ionization and Breakdown

Nearly all types of gas switches are operated on the basic principle of ionization and breakdown of gases. It is therefore essential to have some basic understanding on the subject. In a gas switch, transition from the off (insulating) to the on (conducting) state is a typical example involving gaseous ionization and breakdown. Here ionization, is defined as: transformation of the gas atoms into ions by detaching valence electrons from the former and breakdown is defined as transition of the gas from a highly insulating (high resistivity) state to a highly conducting state caused by ionization. A normal gas is an insulator and it can act as a conductor when it is ionized. Its ability to conduct current is determined largely by the degree of ionization it possesses. A highly ionized neutral plasma can act nearly like a metal. There are various ways to ionize a gas. In switch technology, it is commonly accomplished either by using an external source such as injection of charged particles or ionizing radiation into the gas or via the process of collisions between charged particles (chiefly electrons) and the gas molecules within the gas. In either case, the basic mechanism is similar: detach valence electrons from the gas atoms and transform the neutral gas into a partially ionized plasma. Study of the phenomenon of gas ionization by collisions has been carried out quite extensively in the past. Understanding of the subject, however, is still far from complete, chiefly due to the following reasons.

The process of gaseous ionization and breakdown involves many parameters and interactions and is a problem that needs to be described by using four independent variables (x, y, z, t). Theoretical study of the subject in the past has been mostly qualitative or semi-quantitative in nature[3-5]. A rigorous approach to the problem requires functional solutions of the Boltzmann transport equation or fluid equations which are difficult to obtain in analytic forms. An alternative is to use computer and solve the governing equations numerically. However, such an approach has difficulties in obtaining self-consistency in the result[6]. Most of the existing theories were obtained either from idealized situations or based on simplified models. In the following, we shall discuss some of these theories and the pertinent developments.

Early theoretical work in this field was carried out by J. S. Townsend based on the steady state, one dimensional continuity equations of electrons and positive ions[7]. The results of this work are still widely used at the present time, particularly in the field of high voltage engineering. The basic equations Townsend employed were

$$\frac{d(nv_-)}{dx} = \alpha v_- n \tag{5-2}$$

$$\frac{d(Nv_+)}{dx} = \alpha v_- n \tag{5-3}$$

where n, v_- are the density and velocity of electrons, N, v_+ the corresponding quantities for positive ions, α the Townsend first ionization coefficient, respectively. By using a number of assumptions and setting $J_- = nv_-$ and $J_+ = Nv_+$, the current densities for electrons and positive ions can be obtained from Eqs.(5-2) and (5-3) as follows:

$$J_- = \frac{J_0 e^{\alpha x}}{1 - \gamma(e^{\alpha d} - 1)} \tag{5-4}$$

$$J_+ = \frac{J_0(e^{\alpha d} - e^{\alpha x})}{1 - \gamma(e^{\alpha d} - 1)} \tag{5-5}$$

which show the spatial variations of current densities. However, there is considerable difficulties in interpreting the expression given in Eq.(5-5) satisfactorily. Since the mobility of electrons is about two orders of magnitude greater than that of the positive ions, in order for the two current densities J_- and J_+ shown in Eqs.(5-4) and (5-5) having the same order of magnitude, it would require $N \gg n$. However, if it is so, according to Maxwell equations

$$\frac{dE}{dx} = \frac{e}{\epsilon_0}(N - n)$$

the electric field E would then become very not uniform which is inconsistent with the condition of uniform field for dark discharge. If, on the other hand, one completely ignores J_+, then the total current density $J = J_+ + J_-$ cannot be a constant which is essential for the current to be continuous. In other words, in either case there is a difficulty to reconcile the inconsistency. This may be the reason that in most publications only the expression of J_- has been fully discussed and few explanations, if there is any, has been given for J_+.

By multiplying Eq.(5-4) with the electrode area and setting $x = d$, the well known Townsend current formula can be obtained as follows:

$$i = \frac{i_0 e^{\alpha d}}{1 - \gamma(e^{\alpha d} - 1)} \tag{5-6}$$

which represents the total current in the circuit and it has been experimentally verified to be quite accurate under certain conditions. The breakdown voltage V_b was obtained by combining the Townsend breakdown criterion

$$1 - \gamma(e^{\alpha d} - 1) = 0 \tag{5-7}$$

with the empirical relation

$$\alpha = AP \exp\left(1 - \frac{BPd}{V}\right) \tag{5-8}$$

The result is an analytic expression for the Paschen law, namely

$$V_b = BPd \left/ \ln\left\{\frac{APd}{\ln(1 + 1/\gamma)}\right\}\right. \tag{5-9}$$

This expression has been experimentally tested to be valid under certain conditions. However, it is not free from shortcomings. In an examination of Eq.(5-9), one can see that for the values of $Pd < \ln(1/\gamma + 1)/A$, V_b becomes negative which is obviously physically unreasonable. Besides, experimental results agree well with the predictions of Eq.(5-9) only for the values of $Pd \leq 200$ mmHg·cm, at large values of Pd, no such agreement was found[8].

At this point, one may wonder how valid is the Townsend theory. The answer is that it is valid only in a limited sense. The main cause is due to a number of flaws in its basic formulation. Firstly, the basic equations given in Eqs. (5-2) and (5-3) are steady state equations which cannot deal with transient event such as a breakdown. Secondly, Eqs.(5-2)and (5-3) involve four unknowns (n, v_-, N, v_+) normally four equations are required to uniquely determine these unknowns. The results given in Eqs.(5-4) and (5-5) were obtained by using two equations only, therefore in Eq.(5-5) there is an undetermined parameter J_0 whose physical meaning cannot be properly interpreted there. Thirdly, both α in Eqs.(5-7) and (5-8) are supposed to represent the same physical quantity. However, when $d \to 0$, in Eq.(5-7) $\alpha \rightarrow \infty$ whereas in Eq.(5-8) $\alpha \rightarrow AP$ which cannot be reconciled. These flaws are believed to be the major reasons that cause limitations in the Townsend theory.

As the Townsend theory has considerable limitations, a new theory–the streamer theory was proposed independently by several authors[8-10]. Though the theories proposed by different authors were not the same, the basic principle used by these authors was essentially identical. The basic idea was that at a certain stage in the development of a single avalanche, photoionization of the gas in the gap became the predominant process of producing secondary electrons. As a consequence, secondary avalanches were created in the vicinity of the parent avalanche. At a critical stage, the avalanches transform into a streamer which led to the eventual breakdown of the gas. Raether[9] defined the critical stage as that when the charge within the avalanche head reaches a critical value

$$n_c = n_0 \exp(\alpha x_c) \qquad (5\text{-}10)$$

where $\alpha x_c \approx (18 \sim 20)$ and x_c is the average position of the avalanche. Meek[8] on the other hand, chose the condition $E = E_r$ as the critical condition, where E is the applied field and E_r is the radial field at the avalanche head and has the following functional form in V/cm

$$E_r = 5.3 \times 10^{-7} \frac{\alpha e^{\alpha x}}{(x/p)^{1/2}} \qquad (5\text{-}11)$$

In Eq.(5-11), x is the distance which the avalanche has progressed and P is the gas pressure.

Streamer theory has been experimentally tested and was found to be generally consistent between theory and observations at large values of Pd. It is particularly successful in explaining the phenomenon of filamentary breakdown of gases. However, as in the Townsend theory, streamer theory also has limitations. The main difficulty associated with it is that it is incapable of achieving self-consistency in its result. The major causes are again fundamental in nature and they are embodied in the basic formulation. From Eq.(5-11), one can see that as $x \rightarrow 0$, $E_r \rightarrow \infty$, which implies that before the avalanche has progressed any substantial distance, the field at the avalanche head has already reached a level that is exceedingly high. That is obviously unreasonable.

In addition to the two theories discussed above, there has been a number of other theoretical models developed in recent years. The notable ones are the avalanche chain model[11], the two-electron model[12] and the 3-D analytic model[13] etc. Except for the 3-D analytic model, all the other models have similar difficulty as that of the Townsend theory and streamer theory, namely they are incapable of achieving self-consistency in the results. The 3-D analytic model mentioned above is virtually the only analytic model that can achieve total self-consistency in its results. Other features in the model are that it is formulated in three dimensional form and can be applied for wide range of Pd values. These features make the 3-D analytic model unique. The main limitation of the model is that it is insufficient to deal with a transient event such as breakdown, as it was formulated on the basis of quasi-steady state equations. However, the analytic expression for the Paschen law derived from it, namely[14]

$$V_b = \frac{\theta^2}{A} \left\{ \frac{D_l}{\mu} - \frac{D_l'}{\mu'} \right\} \frac{1}{Pd} + BPd \qquad (5\text{-}12)$$

has a number of advantages in comparison with the expression given in Eq.(5-9). Firstly, there is no limitation for the values of Pd, i.e. all values of Pd in the range $0 < Pd < \infty$ are applicable. Secondly, expression in Eq.(5-12) clearly demonstrated that the basic process governing the Paschen law is contributed from both the drift (via parameters A and B) and the diffusion (via parameters D_l, μ, D_l' and μ') mechanisms, whereas in Eq.(5-9) the diffusion contribution was completely ignored.

In summary, each of the theories presented above is capable of explaining the phenomenon of gaseous ionization and breakdown only in a limited sense. Application of any of these theories is subjected to specific conditions. There is, at the present time, no single theory that is adequate to account for all the observed phenomena concerning gaseous ionization and breakdown in the broad sense. For instance, the streamer theory is generally adequate in the cases of filamentary breakdown of gases, however it has considerable difficulties to achieve self-consistency in other aspects. The 3-D analytic model is self-consistent but is inadequate to account for the transient event in a quantitative way. The present state of affairs is that, until a comprehensive theory is developed, theoretical approach to the design and operation of gas switches has to be supplemented with empirical methods and scaling laws. Existing theories alone cannot offer a complete solution to the problems involved.

(b) Gas Spark Gaps

Gas spark gaps are considered to be one of the most widely used devices in high power pulse applications. They are relatively simple to build and easy to operate. Their application range is wide. Gas spark gaps can conduct current from few ten ampere to multi-megampere and can withstand voltages up to several megavolts. There have been various forms of gas spark gap switches developed in the past. Fig.5-6 shows three typical designs of the commonly used spark gap switches. The basic structure of these switches consists of two main electrodes A and B which are separated by a gap d. The main electrodes are supported by a structure C made of insulating material and with which a closed chamber is formed. The chamber is filled with gas as the insulating medium. The third electrode D is the trigger electrode and its arrangement varies depending on the design. In the trigatron arrangement, Fig.5-6(a), the trigger electrode is embedded in one of the main electrodes and separated from the latter with insulating material E. In the field distortion arrangement, the trigger electrode is located on an equipotential

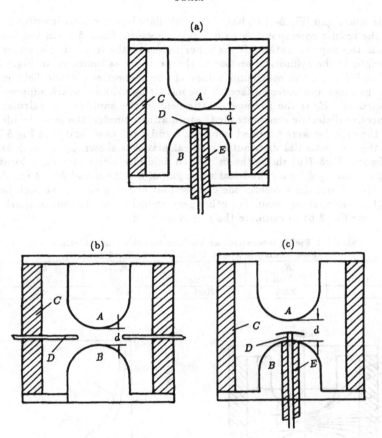

Fig.5-6 Three typical designs of gas spark gaps. (a) Trigatron spark gap.(b) and (c) Field distortion spark gaps. A and B are the current-carrying electrodes. C the supporting and insulating structure. D the trigger electrode.

surface in the field as indicated in Figs.5-6(b) and (c). One important parameter that determines the performance of these switches is the electric field in the gap, as both the processes of ionization during the pre-breakdown stage and avalanche formation that leads to eventual breakdown of the gap are closely related to the electric field. Determination of the field may be done by using the Paschen law given by Eq.(3-4), Eq.(5-9) or Eq.(5-12), if the field is uniform. However, in most actual cases, fields are not uniform, approximate approach in conjunction with empirical data are frequently followed. To assist with the design of gas spark gaps, electric fields correspond to various geometrical dimensions of a cylindrical two-electrode spark gap [Fig.5-7(a)] have been calculated by previous investigators[1]. One of the results corresponding to $h = 0$ is shown in Table 5-1. In the table, d represents the gap separation, R the inner radius of the insulating cylinder and h the length of the cylindrical section of the electrode as indicated in Fig.5-7(a). E_g, E_i and E_0 are the maximum values of the respective electric field in the gap, on the inner and outer surfaces of the insulating cylinder which supports the two electrodes. E_a is the average electric field in the gap. In the calculations, the respective dielectric constants for the insulating cylinder, the gases inside and outside the cylinder were assumed to be $\epsilon = 3$ and $\epsilon = 1$ as indicated in Fig.5-7(a). One of the equipotential distributions so calculated is shown by a rough sketch in the figure. Fig5-7(b) shows the calculated field intensities at various locations versus gap spacing d for a cylindrical spark gap of $h = 2$ cm and $R = 4$ cm shown in Fig5-7(a). From these results one can see that one way to achieve high field is to keep the gap spacing small. For other geometrical configurations of spark gap, one may use Eq.(3-6) to estimate the maximum field.

Table 5-1 Fields intensities at Various locations in a Cylindrical,
Two-electrode Spark Gap of $h = 0$, $R = 4$ cm [1]

d (cm)	E_g (kV/cm)	E_a (kV/cm)	E_i (kV/cm)	E_0 (kV/cm)
0.5	2090	2000	403	422

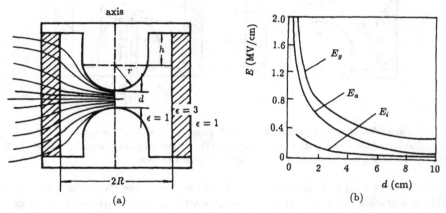

(a)   (b)

Fig.5-7 (a) a sketch (not in scale) of the equipotential surfaces in a two-electrodes spark gap of cylindrical form; (b) fields variation with d for $h = 2$ cm, $R = 4$ cm.

A. Self-breakdown gaps.

In application, the switches shown in Fig.5-6 can be operated either in self-breakdown mode or in a mode that breakdown is initiated by some external source. In the self-breakdown mode, spark gap breaks down when the voltage across the gap is increased above the hold-off voltage. If the increase of voltage is carried out slowly, the breakdown is referred to as static breakdown. If the self-breakdown is caused by a transient voltage, it is called pulsed self-breakdown. For static self-breakdown, the electric field in the gap may be determined from the Paschen law if the field is uniform. For non-uniform field cases, one can use the uniform field data obtained from the Paschen law as an initial base and apply scaling law or use empirical data to refine the design so that the required field is satisfied. For pulsed self-breakdown, the breakdown voltage can be significantly higher than the static breakdown voltage if the risetime of the impulse voltage is sufficiently short as already mentioned in Section 3-2(b). This is because the effect of electron production by ionization and subsequent streamer formation is determined by both time duration and field strength. If the risetime of the impulse voltage is short, a higher field is required to compensate the short time duration in order to produce the same effect for breakdown.

As for designing of self-breakdown spark gaps is concerned, the main aspects need to be considered are the temporal behavior of the applied voltage, the electrode configuration, electrode material, surface condition and the properties of the dielectric material etc. At the present time, there is no available formula that can reliably describe the relationship between all these parameters. The only available source in this respect is the empirical relation established by J. C. Martin under certain conditions as described in Section 3-2(b) by Eq.(3-16). Other sources of more recent information may be found in Ref.[15] in which the author has reported various results concerning pulsed breakdown.

B. Electrically triggered gaps

If the breakdown is initiated by an external source, the switch is said to be operated in triggered mode. There are various ways to trigger a switch. The most widely used method perhaps is the technique using the trigatron scheme shown in Fig.5-6(a). In operation, the main gap is charged to a voltage slightly less than the static hold-off voltage. Breakdown of the main gap is initiated by the application of a fast rising electrical pulse to the trigger electrode. The breakdown process is believed to be physically initiated by streamers launched from the tip of the trigger electrode[16]. After a delay of typically a few tens of nanoseconds, breakdown between the two main electrodes takes place. The trigatron switch can be used in a variety of applications under a wide range of conditions. There are a number of advantages which the trigatron switch can offer. They include:

(1) Wide triggering range – the main gap can be readily triggered at charging voltages as low as 25% of the static hold-off voltage. The delay and jitter, however, will be rather poor at these low voltages.

(2) Good triggering ability – it produces relatively short delay and jitter[17].

For a gap of 100 kV hold-off voltage, a delay in 20–30 ns range with jitter less than 5 ns can be achieved.

(3) Simple structure – trigatron has a relatively simple structure. In operation, only a voltage pulse is required to apply on the trigger electrode.

The main disadvantages associated with the application of it are:

(1) Trigger generator requirement – it requires a powerful enough generator to produce a fast rising, voltage pulse of magnitude comparable to the main charging voltage meanwhile it provides a current of modest intensity.

(2) Electrical isolation of trigger – the trigger circuit is not isolated from the main gap circuit and potential on the trigger electrode may rise to the main charging voltage during switch closing. Some measure is necessary to protect the trigger generator such as connecting a series resistance to the trigger circuit. However, such a measure may affect the switch's performance.

(3) Trigger electrode erosion – in high current operations, erosion of the tip of the trigger electrode will adversely affect the performance of the trigger mechanism.

The electrical trigger method has also been widely used in three-electrode field distortion spark gaps as shown in Fig.5-6(b) and (c). These switches are essentially two gaps in series with the trigger pulse applied to the center electrode which causes one of the two gaps breakdown and subsequently the second gap follows as a result of field enhancement in the main gap. There are two different approaches to create the field enhancement. In one approach the field in a gap which is part of the whole switch, is changed uniformly by the applied trigger voltage resulting in a homogeneous field enhancement. Since the maximum field enhancement so produced is relatively small, generation of initial electrons by field emission mechanism is nearly impossible, external preionization source is required. Besides, there are a number of other disadvantages which make this approach even less attractive[18,19]. The other approach is to use the field distortion produced by the trigger electrode at its sharp edge as a means to trigger the main gap. Such field enhancement is non-uniform and the field strength at the edge of the trigger electrode is high enough to produce field emission electrons. This approach is called field distortion method. The basic idea of the operation is to change the electric field distribution of the spark gap in such a way that the ratio of the maximum field to the average field changes drastically from a value close to 1 in the hold-off state to a value much greater than 1 in the triggered state. The field distortion method of triggering can be used in many types of spark gap switches[2,20], the switches shown in Fig.5-6(b) and (c) are two of them. For more details in this respect one may refer to Ref.[2]. In all designs the main electrodes are designed to generate a field distribution as homogeneous as possible to provide a large hold-off voltage. The trigger electrode is usually with a sharp edge and located on an equipotential surface. The potential of the trigger electrode is kept at a value determined by its position relative to the main electrode. A voltage divider in the trigger circuit is required to eliminate any field enhancement. When a voltage pulse is applied to the trigger, maximum field distortion will occur at the edge of the trigger electrode. There are two different modes of operation, the cascade breakdown and the simultaneous breakdown. In the latter operation, breakdowns of the two gaps occur nearly simultaneously. This type of operation, however, requires specific conditions to be satisfied[2], therefore it is not as widely used as the cascade operation. In the cascade operation, the main voltage is usually positive and a negative voltage pulse is applied to the trigger electrode which initiated breakdown of the first gap (between trigger electrode and anode). Subsequently the second gap breaks down as a result of the trigger electrode potential swings toward the voltage of the anode. The time durations and sequence of these events depend on the capacitance between the trigger electrode and the main electrodes. There is no particular condition required for this type of operation. As for the operation of the gaps shown in Fig.5-6(b) and (c), two examples are given in Table 5-2 which show the various parameters and geometric dimensions involved.

Table 5-2 Operating parameters of the spark gaps shown in Fig.5-6(b) and (c)

Gap design	Gap separation (cm)	Trigger electrode distance ratio	Electrode diameter (cm)	Gas and pressure (atm)	Working voltage (kV)	Trigger voltage (kV)	Delay time (ns)	Jitter (ns)	Max. current (kA)	Ref.
Fig.5-6(b)	0.4	0.5		N_2 up to 10	positive 20–50	negative 20–50	20	2	100	21
Fig.5-6(c)	7.6	12	9.0	SF_6 12	positive up to 3 MV	negative 250	25	0.9		22

C. Electron-beam triggered gaps

Spark gaps can also be triggered by injection of the · e-beam in the inter-electrode region shown in Fig.5-8. When the e-beam is injected into the gap from cathode (foil) to anode, i.e. positive injection, the beam electrons can travel across the entire gap and generate a space charge along their way by impact ionization. The growth rate of this space charge is determined primarily by the properties of the e-beam such as the energy, duration and cross section, as well as by the electric field and the gas employed. If the growth rate of the space charge is rapid enough, secondary ionization processes may become significant which will lead to formation of avalanches and eventually to complete breakdown of the gap. As impact ionization by e-beam can supply a significant amount of initial electrons, the statistical delay time (jitter) of the switch can be significantly reduced. By using this method, some previous investigators[23,24] have obtained delay times of 1 ns in a N_2 gap at $E/P = 28$ V/cm.torr and 15 ± 0.8 ns in a $N_2+SF_6^*$ gap at 1.7 MV hold-off voltage. In the first instance, the e-beam employed was 180 keV, 10 A, 5 ns and having a cross section of 5 cm^2. When the e-beam is injected from the anode (foil in Fig.5-8) toward the cathode, i.e. negative injection, the beam electrons are retarded by the applied field and the distance they can travel in the gap is determined mainly by their energy. Generation of electrons by impact ionization is mainly in the region where the beam electrons have penetrated in. Though similar breakdown phenomenon as that for the positive injection was observed in the gap under negative injection, the physical processes that are responsible for the breakdown phenomenon are not necessarily identical. One clear evidence in supporting this view is that, under the same conditions the delay time is longer for negative injection than for positive injection[23]. Other results obtained from an open shutter photograph for the discharge channels initiated respectively by the positive and negative injections also show distinct differences[25,26] between them. All these suggest that the physical process that leads to breakdown of a gap under the positive injection scheme is not identical to that when the negative injection is used. Details of the physical processes involved are not well understood at the present time. Further study in this respect is needed.

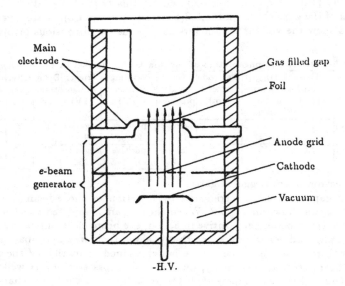

Fig.5-8 Sketch of an *e*-beam triggered gas spark gap.

In the above discussion, the *e*-beam was treated as an external source for triggering the spark gap. The *e*-beam can also be used as an external agent to sustain the discharge in a gap. In contrast to the triggering operation, the sustained discharge ceases to exist when the *e*-beam is removed as a result of electron attachment and electron-ion recombination processes. With this scheme, the on and off states of the switch can be manipulated by properly adjusting the grid voltage shown in Fig.5-8. The *e*-beam controlled diffuse discharge switches are operated on this basis[27,28]. This type of switches can be used either as an open switch or as a repetitive switch. One critical area in determining the performance of the switch is the gas mixture employed which is required to have appropriate attachment and recombination coefficients to satisfy the rapid on and off transitions. With proper choice of the gas mixture, high rep rate greater than 10 kHz, turn on and off time less than 30 ns and at a power level of 200 kV, 20 kA is achievable. One factor that may limit the applicability of this type of switches is the requirement for the e-beam generator. Although the power requirements for the e-beam are modest, the cost of a reliable e-beam generator may limit the application of the e-beam triggered switches to some special applications only.

D. Laser triggered gaps.

The basic process of spark gap triggering essentially consists of three stages. (1) By means of certain external source, some initial electrons are generated in the gap. (2) The growth rate of the space charges is kept fast enough so that secondary ionization processes become significant. (3) The above processes are maintained until a (or more) highly conductive channel is formed. In the process of electrically triggered spark gaps, the first two stages are accomplished mainly by impact ionization of electron neutral particle collisions. As electron energy is derived entirely from the electric field, such an ionization process is inefficient and usually takes a relatively long time to form a highly conductive channel. That means the delay times of those switches triggered electrically are bound to be relatively long. When a laser beam is used for triggering, the situation is different. With high intensity and fast propagation speed, the laser beam can deposit significantly more power per unit volume into the gas, consequently the ionization process can be significantly more efficient. This is the major reason that switches triggered with laser beam usually have shorter delay and jitter in comparison with that for electrically triggered gaps. This feature makes laser triggering particularly attractive for applications where precise timing or synchronization is required. Laser triggering also has other advantages which include:

(1) Electric isolation of trigger-laser trigger circuit can be completely isolated from the high voltage circuit thus enabling the design to be simpler and the operation more safe.

(2) High reliability-good switching performance can be obtained with a working voltage significantly less than the hold-off voltage. This practically eliminates the possibility of prefiring the switch.

The main disadvantage in using laser triggering is the requirement that a laser unit is needed. Unless the cost of laser unit is significantly reduced, use of laser triggering probably remains for special applications only.

Three different methods have been widely used for laser triggering. They are (1) electrode surface triggering, (2) axial volume triggering, (3) transverse volume triggering. All these methods employ a laser beam to provide ionization of the gas, the difference is the spatial distribution of the ionization. In the first method, the laser beam is focused on the surface of one electrode and produces a small plasma fireball on the struck surface as shown in Fig.5-9. The fireball serves as a seed to initiate a streamer. This streamer, with laser assistance, propagates across the gap and forms a filamentary channel of low conductivity. Upon further interaction with the laser, the channel increases its conductivity until the channel develops into a highly conductive one and then closes the gap completely. This method is effective because the streamer leaves behind an ionization density far greater than the required threshold density for triggering, thus bypassing the early stage of ionization growth completely. By using the method of electrode surface triggering in conjunction with optical fibers to transmit the laser light, many previous investigators have frequently obtained results of subnanosecond jitter and few ns delay time in various applications. For details, one may look up references [29–31].

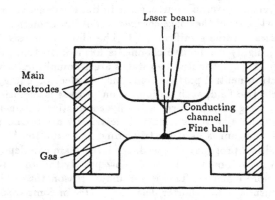

Fig.5-9 Laser triggered spark gap, electrode surface triggering scheme.

In Fig.5-9, if the laser light enters the gap in a collimated form along the gap axis, instead of being focused at one spot on the electrode surface, a thin needle-shaped volume along the laser passage can be highly ionized. Because of the high ionization density in it, this small volume can rapidly develop into a plasma channel of high conductivity and eventually bridges the gap. The method of closing spark gaps in such a way is called axial volume triggering. As intense UV laser beam can generate substantial initial charges along its passage, this scheme has the advantage of reducing the source of delay and jitter in the early stages of channel formation. For this reason, as in the case of electrode surface triggering, this method is equally attractive for those applications involving precise timing or synchronization. On the basis of this scheme, some previous investigators have obtained subnanosecond jitter in a gas switch of 2.8 MV hold-off voltage, triggered with a KrF laser pulse of 40 mJ energy[32]. However, it was found by the same authors, that such results could be obtained only when UV laser was used. If IR laser is employed, the delay time of the switch becomes quite poor[33]. The results may be represented approximately by the two lines shown in Fig.5-10. From the figure, one can see that when the switch is triggered at a working voltage equivalent to 80% of the static breakdown voltage, the delay time for IR triggering is about four times of that for UV triggering. In a recent publication[34], Turman et al.

have reported results concerning the relationship between jitter and gap closure time in a Rimfire switch. The switch is a combination of an axial volume triggered gap with a series of untriggered gaps. By using a KrF laser pulse of 22 ns FWHM with energy between 5 to 60 mJ, the authors have found that the switch jitter increases substantially when the gap closure time becomes longer than the laser pulse width of 22 ns. For more information concerning the subject one may refer to the excellent list of publications given in Refs. [2, 34].

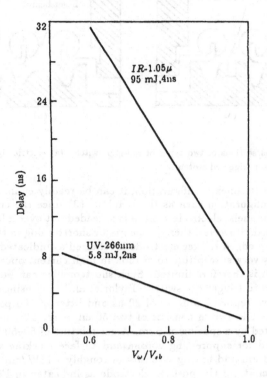

Fig.5-10 Comparison of delay times of a 500 kV switch triggered respectively with UV and IR laser pulses where V_w=working voltage, V_{sb}=static breakdown voltage [33].

The most noticeable feature of the rail-gap switches is that the main electrodes are the two long cylindrical rods or other forms and arranged in parallel configuration. The cross section of two such switches are shown in Fig.5-11 in which (a) represents an electrically triggered switch in conventional form and (b) is a similar design but for laser triggering. The main advantages of the rail-gap switch are (1) it can provide relatively low inductance when it is operated in multichanneling

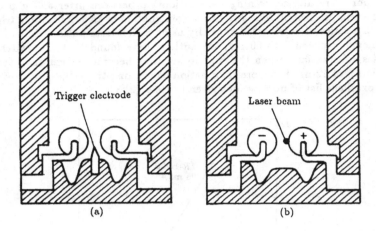

Fig.5-11 Cross sections of two types of rail-gap switch. (a) electrically triggered switch; (b) laser triggered switch.

mode, (2) because its linear configuration, it can be readily connected to systems having similar configuration such as the stripline, (3) since the current is shared by a number of channels, electrode erosion is spreaded out over a large area hence the effect is reduced. However, there is one major shortcoming in the conventional design, i.e. a sharp edged trigger electrode is required as indicated in Fig.5-11(a). The sharp edge is very susceptible to erosion in high current operations, the lifetime of the switch is therefore limited. Such shortcomings can be removed if the laser beam is used to trigger the switch. Taylor et al.[35], by using laser triggering in their work, have reported a delay of 20 ns and jitter of 100 ps obtained from a rail-gap switch. The switch consists of two 50 cm long, 3.2 cm diameter brass electrodes separated by a spacing of 1 cm. A gas mixture of 50–50 N_2 and Ar with 0.075% SF_6 added for suppressing corona and surface tracking was used. KrF laser beam at 248 nm and having intensity of roughly 1 MW/cm^2 or greater was focused at the vicinity of the positive electrode as indicated in Fig.5-11(b). The switch was pulse charged and the operational voltage range was 30–130 kV. As the passage of the laser beam in this type of triggering is parallel to the surface of the electrode, this method is therefore referred to as transverse volume triggering. In a more recent report[36], a 60 cm long rail gap switch filled with SF_6 was triggered with KrF laser of approximately 30 mJ energy and at a working voltage of 500 kV. The delay and jitter were found to be of the order of 50 ns and 2 ns, respectively. It should be noted that, due to its linear geometrical configuration, the rail gap switch does not well fit systems having cylindrical configuration such as the coaxial transmission line.

(c) Multichannel and Low Pressure Gas Switches

A. Multichannel Switches

In high power pulse applications, current pulses of fast risetime are frequently needed. This requires transfer switch of low inductance. Roughly speaking, the inductance of the switch is inversely proportional to the number of conducting channels. This is because the switch inductance consists of many components some of which depend on the number of currents carrying channels N. In this category it is the inductance of the conducting channels L_c, which can be expressed in terms of the radius of the channel a, the radius of the electrode b feeding the channel, the length of the channels d in cm and the number of channels N by[35]

$$L_c = \frac{2d\ln(b/a)}{N} \tag{5-13}$$

Subnanohenry inductance has been achieved with multichannel switch by some previous investigators[37]. There are several types of switches which can operate in the multichanneling mode. The notable ones are the surface discharge switches[38], the rail gap switches[35] and the multi-stage switches[34]. In the following we shall discuss each of these switches respectively.

The basic structures of the surface discharge switches are shown in Fig.5-12. The two main electrodes are wedge shaped and arranged in contact with the dielectric surface or slightly above the surface. The trigger electrode may be placed at various locations depending on the type of design in question. The dielectrics which separate the main electrodes from the ground plane usually consist of two different materials if long lifetime of the switch is required. The upper one generally is made of material having a better ability to resist surface erosion and a lower one is required to have high dielectric strength. The idea is to minimize the effect of surface erosion while keeping good insulation between the main and trigger electrodes. The ground plane is optional, use of it is determined by the applications and other parameters employed. Some surface discharge switches can satisfactorily operate without using the ground plane. Surface discharge switches can be operated either in the self-breakdown mode by overvolting them with a fast-rising voltage pulse or in the triggered mode by applying a fast rising pulse to the trigger electrode. For the former operation, the switches are usually constructed without the inclusion of trigger electrode or ground it to zero voltage. For the latter operation, there are many different triggering schemes available and nearly all of them are operated on the same basic principle of electric field distortion. That is the trigger pulse generates a local field enhancement in the gap which in turn initiates the breakdown process. The various triggering schemes are simply different approaches for achieving the same goal of generating a field enhancement effectively. Basing on the performance with respect to the various parameters of switch delay, channel number, lifetime, hold-off voltage and reliability etc., it appears that the embedded trigger scheme shown in Fig.5-12(a) has more attractions than others. The chief advantage is the total isolation of the trigger electrode from

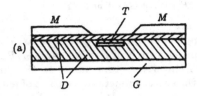

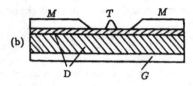

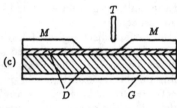

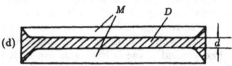

Fig.5-12 Cross sections of three typical types of surface discharge switches: (a) embedded trigger, (b) center trigger, (c) above surface trigger, (d) top view of switch(a). M-main electrode, T-trigger electrode, G-ground plane, D-dielectrics.

the main discharge circuit. By such an arrangement, the main discharge current is prevented from going through the trigger circuit hence increasing the efficiency of the operation. Secondly, it completely eliminates the possibility of trigger electrode erosion thus enhancing the lifetime of the switch. Finally the location and configuration of the trigger electrode can be readily varied to produce the desired effects.

Fig.5-13, shows the basic circuit for the operation of a surface discharge switch with the embedded trigger scheme. A photograph of the top view of this switch is shown in Fig.5-14. The operation was pulse charged with a voltage pulse of 80

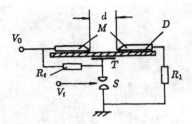

Fig.5-13 Basic circuit for the operation of a surface discharge switch with embedded trigger scheme. V_0-charging peak voltage, R_t-trigger resistance, V_t-trigger pulse, S-trigger generator switch, R_l-load, t_r-charging voltage risetime, l-main electrode length, P-gas pressure, th-dielectric thickness.

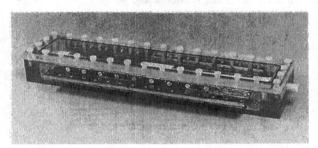

Fig.5-14 Photograph of the surface discharge switch employed in Fig.5-13. The electrodes are 40 cm long [40].

kV peak value and 500 ns risetime[39].

Table 5-3 Parameters employed in the set-up shown in Fig.5-13

Gas	P atm	d cm	l cm	th cm	V_0 kV	t_r ns	V_t kV	R_t kΩ
SF_6+N_2	2	2.5	40	0.4	80	500	15	6

The gas mixture employed was 60% N_2, 40% SF_6 and the dielectric surface used was Boron Nitride with Plexglass. The other parameters employed in the operation are shown in Table 5-3. When the switch was triggered very near the top of the voltage ramp, the channel number was found to be approximately 80 channels/m as shown in Fig.5-15. The switch has been also operated in the self-breakdown mode. The hold-off voltage was found to be strongly dependent on both the impulsing voltage risetime and the gas mixture composition. These results are respectively shown in Fig.5-16 and Fig.5-17. As the performance is concerned, surface discharge switches are in many respects very similar to the laser triggered rail gap switches. They both can provide low inductance switching, can be readily

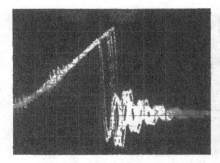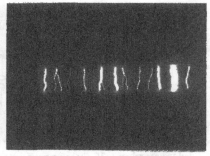

Fig.5-15 (a) Slow pulse charged switch triggered at various points near the top of the voltage pulse; (b) Open shutter photograph of a single shot multichanneling along one pair of the 20 cm long electrodes [40].

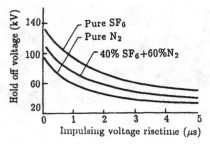

Fig.5-16 Hold-off voltage as a function of gas composition and voltage risetime for the 40 cm switch with $d=2.5$ cm and $P = 2$ atm.

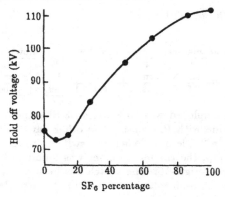

Fig.5-17 Hold-off voltage as a function of SF_6–N_2 gas composition for $d=2.5$ cm, $P=0.2$ Mpa with impulse voltage risetime of 500 ns [39].

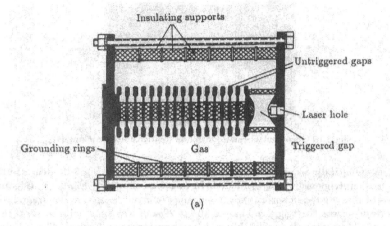

(a)

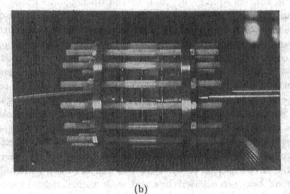

(b)

Fig.5-18 (a)Schematic drawing of the Rimfire switch; (b)Photograph of a four-stage Rimfire switch [42].

connected to systems having linear geometrical configuration and are capable of reducing the effect of electrode erosion. In addition, the surface discharge switch can also serve as a low cost and long life switch for general purpose applications[41]. However, there are also some drawbacks for the surface discharge switches. They are (1) the surface deterioration due to both surface erosion and deposition of foreign materials on the surface, such effects can lead to surface tracking and irregular performance, (2) multichannel switching can be reliably produced only by pulsed charging, it is very difficult to produce the same effect by static charging.

In Subsection (b) of this chapter we have briefly described the Rimfire switch which too is a type of multichannel gas switch. The basic structure of the Rimfire

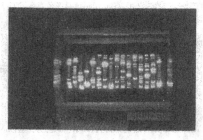

Fig.5-19 Open shutter photo of rimfire multichannel switching [42].

switch is essentially a combination of the switch shown in Fig.5-9 with a number of identical untriggered gaps connected in series as shown in Fig.5-18. Because it consists of many gaps in series, it is sometimes referred to as a multi-stage switch. The voltage across the triggered gap is about 20% of the total voltage and the rest of the voltage is shared equally by each of the untriggered gaps. The maximum field is in the untriggered gaps so that the discharge streamers tend to occur near the rim of the electrodes. In operation, a laser beam first initiates breakdown in the triggered gap and generates a voltage wave moving down the untriggered structure thus strongly over-volts the untriggered gaps. This leads to rapid breakdown of the untriggered gaps. Because the overvolting effect is so strong, several streamers are usually formed independently in each gap hence resulting in multichannel switching. The average number of conducting channels in each gap so formed is about 4 or 5. Some of the channels may leap from rim to rim across several gaps forming a line of discharge. Fig.5-19 is an open shutter photograph of Rimfire switching when it was multichanneling. By employing 36 Rimfire switches in parallel configuration and with each of them standing 6 MV, carrying a current of about 500 kA, investigators of the PBFA-II project have routinely achieved less than 10 ns spread between all switches and with an individual switch jitter of less than 2 ns[43]. In view of the power level involved, such results of synchronization of 36 units are quite impressive.

One other multichannel switch is the gas filled rail-gaps which has been discussed in a preceding section under the subtitle of "Laser Triggered Gaps". As far as the ability of multichanneling is concerned, rail gap switches appear to be modest in comparison with the surface discharge switches and the Rimfire switches. Their attraction may be attributed to their ability of conducting relatively high current while maintaining a moderate number of conducting channels. For example, the commercially available rail gap switch model 40200, manufactured by Maxwell Laboratories, Inc. has a current ability of 750 kA which is considerably better than that of a comparable surface discharge switch operated in multi-channel mode. Other specifications of the Maxwell model 40200 rail gap switch are given in Table 5-4. From this one can see that the requirement for the trigger pulse generator is rather stringent. It is required to generate voltage pulses with $dV/dt \geq 10kV/ns$ and having a peak value of 100 kV. This is one of the reasons that such a type of rail gap switches has not been as widely used as those gas switches discussed in Subsection (b) of this chapter.

Table 5-4 Specifications of Maxwell model 40200 rail gap switch

Voltage Range (kV)	Peak Current (kA)	Charge Transfer (C)	Life Time (Shot)	Inductance (nH)	Trigger Pulse	Jitter (ns)	Pressure Range (PSIG)	Gas
50–100	750	10	5000	20	100 kV rise< 10ns	< 2	12–60	Ar/SF$_6$

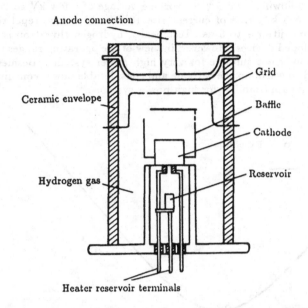

Anode connection

Grid

Ceramic envelope

Baffle

Cathode

Hydrogen gas

Reservoir

Heater reservoir terminals

Fig.5-20 Schematic of the cross-sectional view of a single-gap thyratron.

B. Low Pressure Gas Switches

Hydrogen thyratron is one of the most well developed low pressure gas switches. Many different versions of thyratron have been developed in the past. The basic structure of the single-gap thyratron is shown in Fig.5-20 and the device mainly consists of an anode, a control grid, a baffle, a heated cathode and a hydrogen reservoir. The respective functions of these elements are as follows. The grid is employed to control the number of electrons reaching the anode. A sufficiently intense electrical pulse applied to the grid will initiate rapid ionization of the gas between the grid and the cathode which will subsequently cause the anode-cathode gap to break down. The grid is usually placed very near the anode so that the spacing between them is less than one electron mean-free path at the operating gas pressure. Such a measure is to prevent undesirable breakdown between the anode and the grid. The hot cathode is to supply sufficient electrons in the region between the cathode and the baffle. The function of the baffle, which is maintained at the cathode potential, is to prevent a large number of energetic electrons emitted from the cathode reaching the anode spontaneously which may lead to spurious triggering of the switch. The tube is filled with hydrogen or deuterium gas at the pressure range of about 0.3 to 0.5 torr and a titanium hydride reservoir is usually employed to maintain the required hydrogen pressure.

In pulse power applications, the hydrogen thyratron can be operated approximately in the following ranges[2,44]: working voltage of a few kV to 50 kV, peak current of up to 5 kA, rate of current rise 10^{10} to 10^{11} A/s, repetition rate of up to 10^3 Hz and jitter 1 to 5 ns. Technically, hydrogen thyratron is one of the most well developed devices. However, in view of the operating ranges it can offer, clearly thyratron is not suitable for very high power systems. Besides, the long period required to warm up the cathode and considerable power consumption may further hinder its applicability for high power systems.

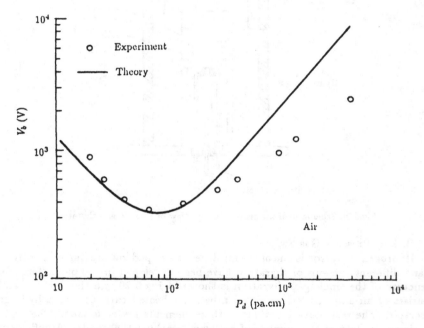

Fig.5-21 Breakdown Voltage V_b versus Pd values for a air gap.

Fig.5-21 is a plot of V_b versus Pd from Eq.(5-12) for air. The curve is generally referred to as the Paschen curve which describes the relationship between

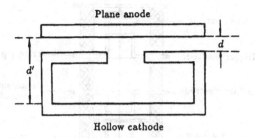

Fig.5-22 Cross sectional view of a discharge set-up with a hollow cathode.

the breakdown voltage V_b and the Pd values for a uniform field gap. Let us now replace the plane cathode in a gap of plane parallel electrode geometry by a hollow cathode as shown in Fig.5-22 and assume that both values of Pd and Pd' are on the Paschen curve but on the left side of the minimum breakdown voltage. Since here $Pd < Pd'$, from the curve we can see that the gap d' will break down at a voltage lower than that of the gap d. This is because in the small gap d, the mean free path of the electron is comparable to the gap spacing and impact ionization due to electron neutral particle collisions is very inefficient, in addition, electron loss due to diffusion in this case is more rapid, therefore higher breakdown voltage is required. In the hollow cathode, the E/P value is low which favors ionization growth, and as a result substantial discharge along the symmetry axis of the hollow cathode can be established with electrons supplied from the cathode region. Positive ions produced in the anode-cathode region are drifting along the symmetry axis into the hollow cathode where they build up a positive ion layer near the cathode. Such discharge can be sustained as long as there is no perturbation to change the balance of electron production and loss rates. If, by some external means, some additional electrons are generated in the cathode region, this sustained discharge will develop into a plasma channel along the symmetry axis and subsequently bridge the anode-cathode gap. Switches operated on this basis are called pseudospark switches.

The phenomenon of pseudospark was discovered in 1979 and since then substantial investigations on the subject have been carried out by many researchers[45]. There have been various types of pseudospark switches developed. On the basis of the trigger method employed, pseudospark switches may be classified into three basic types: electrical pulse triggered[46], surface discharge triggered[47] and UV or laser triggered[48] switches. The cross sectional view of one of the electrical pulse triggered pseudospark switches is shown in Fig.5-23. The switch consists of a pair of specially shaped anode and cathode. The purpose is to prevent the insulator from being metalized by sputtered electrode material. The auxiliary electrode is to provide a weak ionization inside the trigger section prior to the trigger pulse is

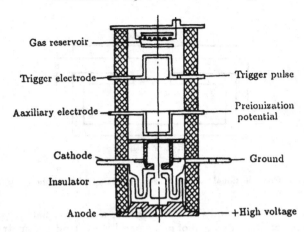

Gas reservoir

Trigger electrode — — Trigger pulse

Aaxiliary electrode — — Preionization potential

Cathode — — Ground

Insulator —

Anode — — +High voltage

Fig.5-23 Cross sectional view of an electrically triggered Pseudospark switch, sealed-off version [46].

applied. When the trigger pulse arrives, such weakly ionized environment can facilitate the initiation of a complete breakdown of the main gap. This is a sealed-off design in which a reservoir is included to maintain the required gas pressure inside the tube. The switch is designed primarily for high repetition rate operation. It can operate at 2 kHz under the condition of 20 kV working voltage and 25 kA current. Other versions of the switch can operate at 0.1 Hz with hold-off voltage of 30 kV and peak current of 100 kA.

Fig.5-24 shows a schematic of the cross sectional view of a surface discharge triggered pseudospark switch. The spacing between the anode and cathode is 5 mm and with a hole of 3 mm diameter at their centers. The surface discharge trigger electrode is embedded between two thin insulating disks which are integrated into the cathode. Upon application of a trigger pulse to the trigger electrode, a sliding discharge occurs along the trigger electrode which presumably initiates the breakdown between the anode and the cathode. The exact nature of the triggering process is not yet fully understood. It appears that the sliding spark also vaporizes parts of the insulator thus causing significant erosion on it. As a consequence the lifetime of the surface discharge trigger is reduced. This is a basic shortcoming of the surface discharge trigger method. The collector electrode shown in the figure is to prevent fast electrons going directly into the anode region which may cause undesirable triggering. By operating the switch at voltages less than 20 kV and current less than 4 kA, a subnanosecond jitter was obtained. However, in view of the overall operating ranges, pseudospark switches can only meet the medium power requirement. They are not suitable for very high power applications.

In addition to the thyratron and pseudospark switches discussed above, there

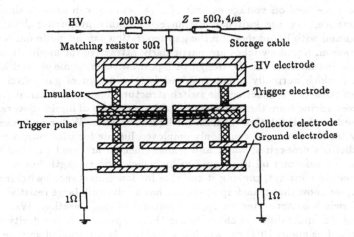

Fig.5-24 Cross sectional view of a surface discharge triggered Pseudospark switch [47].

are a number of other low pressure gas switches which have been also widely used in high power pulse applications. They are the ignitrons[49], vacuum gaps[2], cross field switches[50]. For detailed descriptions of these switches the readers are suggested to refer to the appropriate literatures given at the end of this chapter.

5-3 Liquid and Solid Switches

(a) Liquid Switches

Development of liquid switches is mainly motivated by the high breakdown strength of liquids and the compatibility of the switch structure. Because use of liquid switches is only recent, generally speaking, the technology of liquid switches is not as mature as that of the gas switches. Liquid switches are in many respects quite similar to the gas switches. The main structure and basic elements in the liquid switches are essentially identical to that of the gas switches except the insulating medium. The operating principles of the two are also identical. The differences of the two are exhibited mainly in the performance and behaviors during switching. The different physical properties between liquid and gas are the chief reason that causes such differences. The dielectric strength of most liquids is in the range of 100 to 1000 kV/cm which is significantly higher than that of gas even at high pressure [for more information concerning dielectric strength, one may refer to Eqs.(3-2), (3-4) and Table 3-9 in Chapter 3]. Therefore for switches of the same gap separation, a liquid switch can be operated under much higher hold-off voltage than that for a gas switch. Similarly a liquid switch, operated under the same hold-off voltage as that of a gas switch, can achieve significantly less inductance, because switch inductance is roughly proportional to the gap separation and with liquid medium the gap separation can be made much smaller. For this reason, liquid switches are particularly attractive for applications which require low inductance. In the design of liquid switches, some special attention is essential which normally is not required for the design of gas switches. One example of these is the effect on the switch structure due to mechanical stress and shock waves arising from the high electric field breakdown of liquid. Several liquids can be used as the insulating medium for liquid switches. Water and transformer oil are by far the two most commonly employed liquids for such purpose. Owing to their distinct properties, water and transformer oil are used to serve different functions. Transformer oil is a liquid of high dielectric strength but a relatively small dielectric constant, making it suitable for low inductance switching. From Table 3-1, we know that transformer oil also has a relatively large resistivity, which makes oil switch better suited for applied voltages of long duration. Water, on the other hand, has quite different characters in these aspects. The resistivity of water, when purified, is about $10^7 \Omega$.cm which is suitable only for applied voltage of short duration, otherwise loss through leakage current may become serious. Besides, due to the high dielectric constant ($\epsilon = 80$) of water, capacitance associated with the electrodes may become a significant factor to be concerned with, particularly when the electrodes' areas of the water switch are large and the gap separation is small. Proper care in this respect is needed when designing a water switch of this nature. One advantage of the water medium is its property of self-healing which makes it more suitable for repetitive switching, whereas for transformer oil, it carbonizes during switching and requires some means to remove the carbon residue left in the oil.

In the process of gas discharge, both forms of diffuse and filamentary breakdown are accessible. Breakdown of liquids, however, is generally characterized by luminous channels of filamentary nature. Study of the breakdown phenomenon of the two widely used liquids, water and transformer oil, has been carried out by many investigators. The general understanding on the subject may be summarized as follows[51]. There are always some initial seed electrons present in liquid due to various sources. During the pre-breakdown stage, these initial electrons are accelerated as the applied field increases and at some point they become energetic enough to ionize the liquid molecules. The number of free electrons in the liquid is thus multiplied with the simultaneous dissociation of liquid molecules, forming free hydrogen. The hydrogen bubble then leads to formation of the luminous channels. Quantitative description of the phenomenon in any precise way is very difficult and prediction of the behavior can only be done by approximate methods. The empirical formulas for the prediction of the breakdown voltage under uniform and non-uniform field conditions were given respectively in Eqs.(3-18) and (3-19). For other details concerning the subject one may refer to Subsection 3-3(b). Based on the studies carried out by J. C. Martin and associates, the streamer velocity for point (or edge) to plane electrode geometry in the 100 to 1000 kV/cm range for a few liquids can be expressed approximately by[52]

$$v_s = \frac{d}{t} = kV_b^n \qquad (5\text{-}14)$$

where v_s is in cm/μs, d is the gap spacing, t the time required for the streamer crossing the gap, V_b is the breakdown voltage in MV, and the values of k and n are given in Table 5-5.

Table 5-5 Numerical values of k and n

Liquid	k^+	n^+	k^-	n^-
Water	9	0.60	16	1.09
Oil	90	1.75	31	1.28
Glycerine	41	0.55	51	1.25
Carbon-tetrachloride	168	1.63	166	1.71

Where + and − indicate the initiation from (+) and (−) electrodes respectively [53].

The rise time t_r of the discharge current is interpreted as contributed from two parts, the resistive phase and the inductive phase in the conducting stage of the switching process. It is expressed as

$$t_r = (\tau_R^2 + \tau_L^2)^{1/2} \qquad (5\text{-}15)$$

where τ_R is the part of current risetime corresponding to the resistive phase, τ_L is that corresponding to the inductive phase. τ_R and τ_L can be further expressed as[54]

$$\tau_R = \frac{232}{(NZE^4)^{1/3}} \qquad (5\text{-}16)$$

$$\tau_L = \frac{L}{NZ} \qquad (5\text{-}17)$$

In the above expressions N is the number of luminous channels, Z is the impedance of the driving circuit, E is the mean electric field in the switch and L is the inductance per switch channel. L can be expressed as

$$L = 2d\ln(b/a) \qquad \text{(nanohenries)} \qquad (5\text{-}18)$$

where d is the channel length in cm, a is the radius of the channel and b is the radius of the disc feeding the channel. The rms value of the jitter in MKS units is approximately

$$\delta = 0.825td \qquad (5\text{-}19)$$

where t is the effective charging time during which the applied voltage is greater than 63% of its value at breakdown. From these expressions, one can see that when a liquid switch is operated in the multichanneling mode, a significantly fast risetime of the discharge current may be obtained.

In the design of gas switches, proper measures to prevent surface flashover along the dielectric surface is one of the essential considerations. This exercise is particularly important for the design and operation of liquid switches. Breakdown along the solid-liquid interfaces could cause serious problems to the switching operation. In principle, the hold-off voltage across the solid-liquid interface surfaces is at least as high as that required to break down the dielectric medium. However, this is true only for ideal situations. If any of the following conditions exists, premature breakdown along the liquid-solid surfaces is likely to occur. (1) When there are air bubbles or debris present on the solid surfaces or at the triple-point (i.e. the juncture of the solid-liquid surface with the metallic electrode surface). (2) When the solid surface is contaminated with dirt or has deep scratches. (3) When the bulk material of the solid dielectric contains voids and other imperfections. These foregoing elements constitute the least controllable and easily overlooked factors yet which may cause serious problems for the design and operation of the liquid switches, and therefore proper care must be taken.

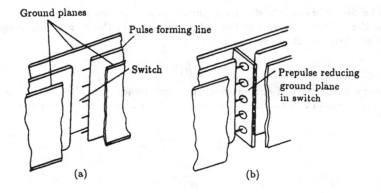

Fig.5-25 (a) Conventional water switch system. (b) Switching system with ground plane. To reduce the effect of prepulse [54].

Like the gas switches, liquid switches can be operated either in self-breakdown mode or in triggered breakdown mode. Fig.5-25 shows one of the water switches operated in the self-breakdown mode. The design is a multichannel switch[54]. The entire system which includes the switch, the pulse forming line and the output transmission line uses water as the insulating medium. Such a design greatly simplifies the structure of system. If gas is used for the switch, many interfaces and diaphragms would be required to separate the gas from the liquid. The ground plane shown in Fig. 5-25(b) is to reduce the effect of prepulse voltage arising from capacitive coupling between the pulse forming line and the output transmission line. Such a prepulse voltage can cause serious problems to the operation of the system. Fig.5-26 is a schematic drawing of a triggered water switch. The entire system consists of a SF_6 trigatron receiving the trigger signal and a main water switch to conduct current. Before triggering, the powered electrode is charged to a potential of $-V$. The disc-shaped trigger electrode, the $CuSO_4$ resistor and the sphere-shaped electrode of the trigatron are charged up to $-V/6$ by means of capacitive division. Upon application of a $+V_t$ trigger pulse to the trigger pin, the SF_6 trigatron switch closes thus changing the potential of the disk electrode to that of the nearby grounded electrode. As the potential of the disc electrode changes, the field enhancement at its edge initiates streamers from the edge toward the negative electrode. When these streamers reach the negative electrode, the water switch closes. At this point, the potential of disc electrode will change rapidly toward the potential of the negative electrode. The $CuSO_4$ resistor here is to serve the function of decoupling the trigger disc from the switch connection to the positive electrode.

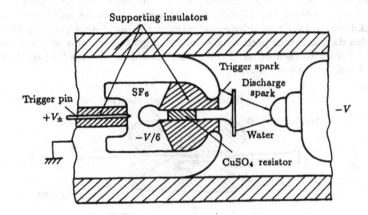

Fig.5-26 Schematic drawing of a triggered, multichannel water switch [55].

Both examples illustrated above are water switches. Many types of liquid switches using transformer oil as the insulating medium have also been developed. One such example is the triggered 12 MV oil switch for the Aurora accelerator[56]. In a recent study[57], liquid switches using flowing transformer oil have demonstrated capability of hold-off voltages up to 290 kV and operated at a repetition rate of 200 Hz. Detailed discussion of these switches is deferred until a latter section when the subject of high repetition rate switches is discussed.

(b) Solid Switches

A. Solid Dielectric Gaps

Development of solid switches may be considered as the extension of the development of liquid switches. As the breakdown strength of most solid dielectrics is at least one order of magnitude greater than that of the liquids, by inserting a thin sheet of solid insulator between two electrodes, one can achieve the same hold-off voltage as that of a liquid switch, yet using much smaller gap spacing. From Eqs.(5-18) and (5-19) one can see that small gap spacing means low inductance and low jitter for the switch performance. This is the main motivation for the development of solid dielectric switches. Solid switches, however, are not free of shortcomings. One of the major obstacles that prevents them becoming popular devices is that solid dielectrics are not self-healing and cannot be reused after switching whereas gas and liquid generally do not have such problems. For this reason, solid switches are most suitable for the cases where high current and single-shot operations are required. The behaviors of the solid dielectric switches are governed mainly by the breakdown process of solid dielectrics. Prediction of the behaviors in any precise way, however, is very difficult chiefly because the subject is still not well understood. Though some empirical formulas for the prediction of the breakdown voltage and substantial numerical results were given in Subsection 3.4(a), this data is rather preliminary and incomplete and some of them may even be contradictory to each other. Proper care should be observed when they are to be used for accurate evaluations.

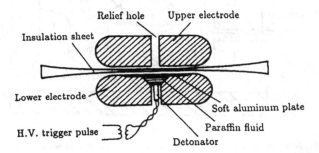

Fig.5-27 Cross section of a detonator triggered, solid dielectric switch [58].

On the basis of breakdown mode, solid dielectric gap switches may be classified as self-breakdown and triggered breakdown switches. The basic structure and operating principle of the self-breakdown switches are quite similar to that of the gas switches, therefore discussion of them is omitted here. There are several types of triggered solid gap switches. Fig.5-27 shows the cross section of a detonator triggered solid gap switch[58]. In the design, the upper electrode has a hole with a diameter roughly equal to that of the detonator. An insulation sheet is placed between the two main electrodes. The detonator is embedded in the lower electrode with some paraffin fluid packed on top of it. The action of closing the main gap is initiated by the explosion of the detonator when it is triggered externally. The high explosive pressure punches through the soft aluminum plate and the insulation sheet, nearly meanwhile a high voltage arc is formed and subsequently released into the in-line hole of the upper electrode. The gap is thus closed. In Fig.5-27, if the detonator is replaced by a piece of wire or foil which is confined in a closed volume, then it can serve the same function of triggering as that of the detonator. The triggering action in this case is initiated by the rapid resistive heating of the confined foil. Upon such heating, the foil vaporizes quickly and the high pressure generated in the process will puncture the soft aluminum plate and the insulation sheet thus bridging the two electrodes with a high voltage arc[59]. A third approach of triggering the solid gap switches is the metal-to-metal method[59] in which a folded metal plate, with one end connected to the lower electrode, is placed below the insulation sheet. Upon exploding the foil, which is inserted in the folded metal plate, the folded metal plate is rapidly displaced toward the upper electrode, so that it cuts the solid insulation sheet and becomes embedded in the hole of the upper electrode. The two main electrodes are thus bridged by the deformed metal plate.

All these methods described above for triggering solid dielectric gaps have been employed in high power pulse applications. The operating parameters of these switches fall approximately in the range of a few to 250 kV hold-off voltages, 100–1000 kA discharge currents and a few ns to several μs current risetimes. Based on the descriptions and results presented in the previous reports, it appears that, among the three types of solid dielectric gap switches, the one with detonator triggering is superior to the other two[60]. One of the reasons is that it is easy to operate and simple to construct.

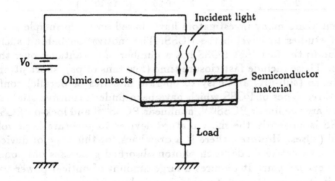

Fig. 5-28 Schematic illustration showing photoconductive semiconductor switch in a simple circuit.

B. Photoconductive Semiconductor Switches

Photoconductive Semiconductor switch (PCSS) is a type of solid switch working on a basic principle rather similar to that for the laser triggered gas spark gaps described in Subsection 5-2(b), namely the switch medium is converted from its initial state of insulation to a highly conductive one by means of light pulse. A PCSS consists of basically a piece of semiconductor material as the insulating medium with two ohmic contacts serving as the electrodes as shown in Fig.5-28. The current density going through the semiconductor (assuming intrinsic) is proportional to the density of conduction electrons in it. Under normal conditions, most of the electrons are localized with the host atoms of the semiconductor and only a small number of electrons are capable of conducting current, therefore the conductivity of semiconductor is quite low $(10^{-5}-10^{-1}\Omega\cdot\text{m}^{-1})$. When an incident light of photon energy greater than the band gap energy falls on the surface of the semiconductor, some of the localized electrons will become mobile and be capable of conducting current. Generally speaking, the current density is proportional to the intensity of the incident light. Therefore, one can use an incident light pulse as a trigger to quickly convert the semiconductor from a nearly insulating state to a highly conductive one. Several semiconductor materials which are potentially suitable for the development of PCSS and their basic properties are listed in Table 5-6.

Table 5-6 Potentially suitable materials for PCSS

Material	Breakdown Field (kV/cm)	Band Gap Energy (eV)	Maximum wavelength (μm)	Intrinsic Carrier Density (cm^{-3})	Maximum Current Density (kA/cm^2)	Electron Mobility (cm^2/V.S)
Ge	80	0.67	1.85	2.4×10^{13}		3900
Si	300	1.11	1.09	1.4×10^{10}	160	1350
GaAs	350	1.4	0.89	9×10^{8}	320	4000
InP	250	1.3	0.95	1×10^{8}		4600

In recent years many investigators have, based on the principle stated above, carried out studies to develop the PCSS. The motivation behind such developments is due to the fact that PCSS has a number of advantages over the conventional ones. These include fast closure time (few tens of ps), short jitter (few ps), low inductance (sub-nH) and easy to operate in series or parallel configuration. There are three types of PCSS which have been under extensive studies in recent years. They are the linear PCSS[61], nonlinear PCSS[62] and lock-on PCSS[64]. The linear PCSS is currently the most mature devices to generate high voltage subnanosecond pulses. However, there is a drawback for this type of devices. In the linear photoconductive mode, each photon absorbed generates only one electron-hole charge carrier pair. It requires a large amount of optical power to maintain the switch in its conducting state, therefore making it impractical for many applications. To date, their application has been limited to those systems in which the closure time, jitter and inductance could not be obtained in any other way. By using a Nd:YAG laser pulse of 35 ps width, Giorgi et al. have, in two experiments, obtained output power of 118 MW with a pulse risetime of 32 ps, and peak current of 6.4 kA in 4 ns from a Si linear PCSS[61].

In the non-linear PCSS switch, closure can be initiated with much less optical power than that required by the linear PCSS. But the required electric field to operate the former is large. For example, before switch closes, the switch must operate at an electric field of about 100 kV/cm. After closure, the electric field strength must be about 10 kV/cm to maintain the carrier density during conduction. Such a phenomenon was previously interpreted as due to an avalanche process similar to that in the gas breakdown where the electron multiplied rapidly once the process is initiated. Experimental results obtained, however, cannot be readily explained with such an avalanche model, because the closure times observed were much slower than predicted by the model. It is likely that the observed phenomenon is contributed from several processes in which the avalanche process possibly is one of them. Many investigators have reported a wide variety of results concerning non-linear PCSS. However, the basic physical mechanism which is responsible for the observed phenomenon is still not well understood. Some preliminary results obtained from a GaAs PCSS indicate that at an electric field of 28.8 kV/cm, switch closure time less than 1 ns is obtainable[63]. It appears that substantial studies are needed before the non-linear PCSS can become a mature technology.

The main features of the lock-on PCSS are:

(1) At high fields (about 4–12 kV/cm depending on the type of material employed), both GaAs and InP exhibit lock-on phenomenon, i.e. when they are optically triggered, they do not recover as long as they are holding an electric field greater than the threshold intensity (about 4–8 kV/cm). At this point, the switch will carry as much current as a circuit can supply.

(2) The optical energy required to trigger a switch into lock-on is much less (about 1/500) than that required to activate the same switch at low fields (i.e. linear photoconductivity).

(3) The resistance of the switch recovers when the electric field across it is reduced to a value below the threshold.

By using a Cr:GaAs switch of dimensions 1.5 cm long, 2 cm wide and triggered with a 532 nm laser pulse of peak power 7 MW, 10 ns wide, Zutavern et al. have obtained the following results[64]. The switch was pulse charged to −32 kV across the 1.5 cm dimension. When the laser fired, the voltage dropped to +12 kV and the current increased from zero to 400 A. These levels remained relatively constant for about 160 ns until the stored energy had been dumped from the charging system then the voltage and current dropped to zero again. The highest current achieved in a lock-on PCSS is 4.2 kA which was obtained from a 0.2×3.0 cm^2 GaAs switch charged to 4.6 kV. The switch was triggered with a 830 nm pulse of 100 ns wide and over 200 μJ energy[65]. In the work, current filamentation and damage caused near the contacts have been observed. These effects are likely to become serious when long term application of the switch is considered. In summary, the main attractions of the PCSS are their fast closure and low jitter times, however, their limited operating ranges of voltage and current make them unsuitable for high power applications.

5-4 Plasma Switches

Strictly speaking, nearly all types of high power switches can be considered as a plasma switch. From Fig.5-3, we can see that during the switching process there is always some form of plasma involved. Those switches described in the preceding sections were so classified mainly based on the nature of their initial state. The switches we are referring to in this section are those switches whose initial state involves some form of plasma. On this basis, the following switches are classified as plasma switches.

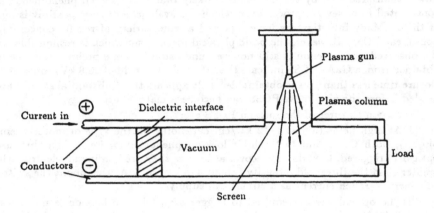

Fig. 5-29 Cross-sectional schematic of a plasma-erosion-switch system.

(a) Plasma Erosion Switch

Plasma erosion switch (PEOS) is a relatively newer type of switch whose main functions can either be working as an open switch in an inductive energy storage system to produce fast risetime pulse for the improvement of output power level or working in a conventional pulse generator to reduce the prepulse effects. In either case, the switch is required to work in an environment having proper vacuum condition. The basic set-up for an operational plasma-erosion-switch system[66] is shown in Fig.5-29. In the operation, carbon plasma from the plasma gun is injected toward the negatively stressed conductor so that the switch region is filled with highly conductive plasma that, in effect, is equivalent to connect the two conductors with a conducting medium in parallel with the load. A short time later, the generator is fired and the voltage pulse arrives at the plasma region which leads to a current flowing through the plasma. As the PEOS is a short circuit and the load is isolated from the generator, at this stage the electric energy is stored inductively in the inductor L as shown in Fig.5-30. The electron and ion currents are taken to be related by the bipolar space charge limited flow as

$$I_e = I_i(Zm_e/M_i)^{-1/2} \tag{5-20}$$

with

$$I_i = 2\pi r Z e n_i v_d l_c(t) \tag{5-21}$$

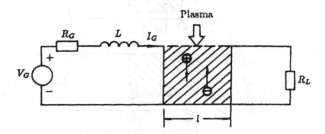

Fig.5-30, Equivalent circuit of the plasma-erosion-switch system shown in Fig.5-29.

where m_e and M_i are respectively the mass of electrons and ions, Z the ion charge state, r the radius of the inner conductor, e the electronic charge, n_i the ion density in the plasma, and assumed to be uniform and constant, v_d the ion velocity and is assumed to be equal to the plasma injected speed and $l_c(t)$ the eroded plasma width as shown in Fig.5-31(a). For C^+ ions, the total current is 99% contributed from electrons, however, the 1% ion current is physically important because it determines how the erosion process changes with time. The switch current $I_s = I_G - I_L = I_e + I_i$ in this case is approximated as

$$I_s \approx I_e = [M_i/(Zm_e)]^{1/2} 2\pi r Z e n_i v_d l_c(t) \tag{5-22}$$

where I_G and I_L are respectively the total current from the generator and the current going through the load. As the switch current increases so does the quantity $l_c(t)$, at $l_c = l$, the switch current becomes

$$I_s \approx [M_i/(Zm_e)]^{1/2} n_i e Z v_d 2\pi r l \tag{5-23}$$

At this point, conduction phase ends and the switch voltage is described by the Child-Langmuir law as

$$V_s^{3/2} = 3.7 \times 10^4 I_s D^2/(rl) \tag{5-24}$$

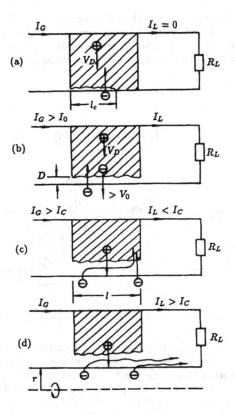

Fig.5-31 Four phases of PEOS operation: (a)conduction; (b)erosion; (c)enhanced erosion; (d) magnetic insulation [67].

where D is the gap width of the eroded plasma as shown in Fig.5-31(b).

The above process is referred to as the conduction phase. As the process proceeds, more ions than electrons enter the eroded plasma because the Child-Langmuir law limits the number of electrons entering the neutral plasma. As a consequence, the gap D increases with time and meanwhile the ions gain an additional velocity component dD/dt, thus

$$I_i = Zen_i 2\pi rl(v_d + \frac{dD}{dt}) \tag{5-25}$$

This process is termed as erosion phase and during this phase current may begin to flow in the load.

When the switch current reaches a critical value Ic i.e.

$$I_s = I_c = 1.36 \times 10^4 (\gamma^2 - 1)^{1/2} r/D \qquad (5\text{-}26)$$

magnetic effect becomes important which forces the electrons emitted from the cathode surface to bend their trajectories toward the load as shown in Fig.5-31(c). This leads to a further increase of the current in the load and more rapid growth of D. Enhanced erosion is named for this phase.

When $I_L \geq I_c$, the magnetic field blocks the electrons completely from traversing the gap and the switch is said to be magnetically insulated. This phase is referred to as a magnetic insulation phase. At this point, the switch is open and nearly all the currant is flowing in the load except a small amount (typically 1% of I_G) of ion current remaining in the switch. This phase is illustrated by Fig.5-31(d).

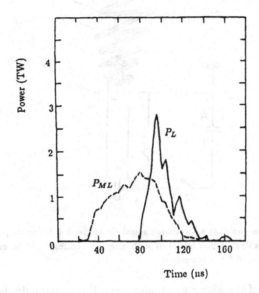

Fig.5-32 Gamble II results obtained with application of PEOS, load power P_L compared with matched load power P_{ML} [67].

The above interpretation has been successfully applied to predict experimental results under a large variety of experimental conditions[67]. Fig.5-32 is one of the experimental results obtained from the Gamble II generator at NRL in the USA. The load in the experiments was an inverse ion diode to produce deuteron beams. The PEOS conducts current for 50 ns then opens in < 10 ns, transferring 720 kA to the load. The measured currents I_G and I_L are approximately 0.8 MA and 0.75 MA respectively at the peak. The voltage calculated from electrical monitors agrees with the 4.2MV peak load voltage obtained from the nuclear diagnostic. The peak value of the matched load voltage is less than 2MV, showing the pulse compression and voltage multiplication. The corresponding power P_L is shown in Fig.5-32 which has an risetime of 10 ns. The peak power of P_L is about 1.8 times the ideal matched load value P_{ML}, showing the power multiplication obtained through the use of PEOS.

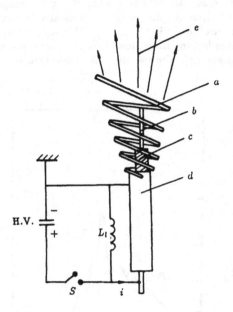

Fig.5-33 Plasma gun (68) a. grounded spiral electrode, 5–6 turns; b. positive electrode (15–20) kV; c. teflon insulator with graphite coating; d. 0.46 cm diameter coaxial tubing; e. C^+ plasma column.

The efficiency of the above mentioned operation is strongly dependent on the PEOS plasma parameters. For a given set of plasma gun, generator, pulse and load, the switch plasma parameters can change drastically by varying any of the following parameters: the plasma gun to electrode separation, the plasma speed, the time interval between the injection of plasma into the switch region and the firing of the pulse generator and the direction of plasma injection with respect to the polarity of the electrodes. Therefore, it is essential to have sufficient knowledge in these respects to achieve optimum operation. It is equally important to be familiar with the basic property of the plasma gun employed. There are two types of plasma guns for this purpose, the carbon gun[68] and flashboard[67]. Fig.5-33 shows the basic structure of the carbon gun and its associated circuit. Measurements[67] indicate that the gun current is an underdamped sinusoid with a peak value of 35 kA and 0.5 μs quarter period and the current profile is sensitive to the spiral form. At $t = 1.5$–2μs after the gun is fired, the peak ion density is about 1–2×10^{13} cm^{-3} and with a drift velocity of 5 -10 cm/μs. The electron density at $t = 2.5$ μs is about 6×10^{13} cm^{-3}. The composition of the plasma was reported to be mainly carbon ions with some neutral H. Results from the flashboard have not been as well studied as the carbon gun, discussion of the results is omitted here.

(b) Plasmadynamic Switch

In the plasma erosion switch discussed in the preceding section, the switching process is accomplished by the action of rapid erosion of the conducting plasma confined in a fixed region which leads to a drastic increase of impedance in the conducting path thus resulting in a transfer of magnetic energy from the inductive store to the load. The plasma-dynamic switch is operated on a similar basis except in this case the switching process is accomplished by using a propagating plasma (as the word plasma-dynamic implies) whose impedance increases drastically when it approaches the load. The basic structure and operating principle of the plasma-dynamic switches are shown in Fig.5-34 which are in many respects similar to that of the well-studied dense plasma focus device[69]. A pair of coaxial electrodes are the main structure and the annular vacuum space between the two electrodes provides the basis for an inductive store when an axially propagating radial discharge is formed between the two coaxial electrodes. There are several ways to create such radial discharge. In this case, a wire array in annular form and backed by a thin dielectric foil is employed to form a short circuit between the two coaxial electrodes. When current flows, the wire-array explodes and transforms itself into a high mass density plasma of an annular sheet. This plasma sheet is axially accelerated upstream by the Lorentz force while the dielectric foil prevents any precursor of wire plasma from spreading downstream. During the acceleration, magnetic energy is accumulated in the coaxial inductive store below the plasma sheet. As the high mass density plasma passes off the end of the inner electrode, the current is switched to flow into the load thus transforming the magnetic energy into it. For a load of cylindrical geometry, if the current is strong and rises fast enough, the load will be imploded onto the axis of symmetry.

In order to provide scaling relations that describe the behavior of the plasmadynamic switch as an element of a circuit, some authors based on a number of simplifications, have derived the analytic expressions for the switch current I, the rate of change of switch inductance dL/dt associated with the motion of the plasma and the optimum switching time t. They are as follows

$$I = \frac{\phi_0}{L_0 + L} \qquad (5\text{-}27)$$

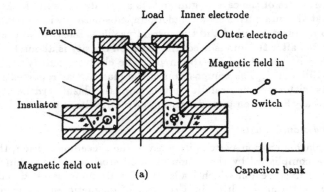

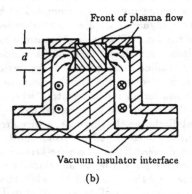

Fig.5-34 Schematic of basic structure and operation of a plasma flow switch. (a) magnetic energy accumulates in coaxial inductive store as annular discharge (wire-array) is accelerated axially toward the load and by Lorentz force; (b) magnetic energy is transferred to the load as high density (mass) plasma formed by the explosion of the wire-array passes off the end of the inner conductor [69].

where L_0 is the storage inductance, I_0 is the initial current in L_0 and $\phi_0 = L_0 I_0 =$ constant. For optimum switching performance

$$\frac{dL}{dt} = \frac{L_l}{\sqrt{m}} \frac{\phi_0}{\sqrt{L_0}} \frac{1}{\sqrt{5}} \qquad (5\text{-}27')$$

If the conduction time is t, then

$$\frac{dL}{dt} = 0.47 \left(\frac{L_0}{t}\right) \qquad (5\text{-}28)$$

where L is the switch inductance, L_l is the inductance per unit length of the coaxial conductors and is constant for a given set-up, and m is the mass of the moving wire conductor. Let us use these expressions to illustrate the procedure for obtaining the relevant parameters of an inductive store system in which the plasma flow switch is employed. Assuming that the inductive store is charged initially with a current $I_0 = 10$ MA, and with total energy $\epsilon_0 = 10$ MJ, $L_l = 2 \times 10^{-7}$ H/m and final velocity $u = 2 \times 10^5$ m/s, since $\epsilon_0 = L_0 I_0^2/2$, we have $L_0 = 2 \times 10^{-7}$ H. From Eq.(5-27) we obtain $\phi_0 = L_0 I_0 = 2$V.S. From $L_l u = dL/dt$, we have $dL/dt = 4 \times 10^{-2}$ H/S and from Eq.(5-28), we obtain $t = 2.4 \times 10^{-6}$ S. From Eq.(5-27'), we obtain $m = (L_l \phi_0)^2/[5L_0(dL/dt)^2] = 0.1$ gm. From these results, we can see that the required mass of the wire conductor is quite small. The switch opening time, for $d = 2$ cm slot, is $t' = d/u = 100$ ns and the switch conduction time is 2.4 μs. As the analytic expressions used to obtain these results are rather crude so are these results, therefore they are meaningful only for estimative purpose. For real designs, the system parameters must be obtained via rigorous approaches.

Cochrane et al.[70], using the plasma flow switch (PFS) have experimentally studied foil implosion. The parameters of the system are as follows: maximum energy of the capacitor bank $\epsilon_0 = 1.5$ MJ, charged at 120 kV, initial charge current into the PFS, $I_0 = 6.5$ MA, current conduction time $t = 3.3\mu$s, plasma sheet velocity (theoretical) u=6–7 cm/μs, switch opening time $t' = 300$ ns, mass of wire-array (aluminum+mylar foil) $m = 40$ mg, dimensions of load (thick wall aluminum cylinder) 2 cm high, 10 cm diameter, electrodes dimensions 7.2 cm i.d. and 10.2 cm o.d., load slot dimensions 2 cm wide and 2.7 deep. The observed result obtained by Cochrane et al.[70] from such a system is shown in Fig.5-35. From the figure, one can see that the risetime of the load current so achieved is quite impressive in view of the current magnitude involved.

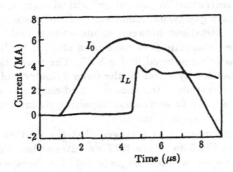

Fig.5-35 Current (I_0) from capacitor bank and current (I_L) switched into a dummy load using a PFS system [70].

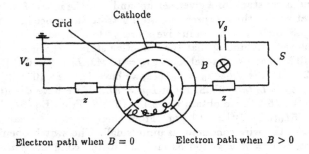

Fig.5-36 Basic structure and operational circuit of triode XFT [71].

(c) Cross Field Tube

Cross field tubes (XFT) are a type of switches that may well be considered as low pressure gas switches, because they employ low pressure gas as the switch medium and operate at low Pd value on the left side of the Paschen breakdown region. We include XFT in this section is mainly on the basis of the reasons stated earlier in the beginning of Section 5.4. Although XFT was invented more than 40 years ago, it is only in recent years, that the technology is maturing to the point where reliable, long life devices are being developed. Two models of XFT have been employed in high power pulse applications. They are the triode XFT and tetrode crossatron modulator switch. The basic features and operating principle of the two are similar except the latter has one more grid electrode. The basic structure and associated circuit of the triode XFT are shown in Fig.5-36. It is usually constructed in coaxial cylindrical form and the tube is filled with low pressure inert gas. Conduction state of the gas is controlled by means of a magnetic field. Relations between anode voltage and gas pressure for Ne and He gases at fixed magnetic field have been studied by Harvey et al.[72] and the result for He gas is reproduced in Fig.5-37. The magnetic field is usually in pulsed form and produced by a coil on the outer cylinder which in general is the cathode. The center electrode is the anode. A grid electrode, whose function is to electrostatically trigger the tube without impairing its interruption capability, is located between the cathode and anode.

In operation, the circuit is so arranged that the grid electrode is positively biased as indicated in Fig.5-36. Since XFT operates at low Pd values outside the Paschen breakdown region, without magnetic field the average electron path is less than the mean-free path and the electrons would be collected by the anode with no ionization produced along their way. However, when a magnetic field of sufficient strength is applied normal to the direction of the electric field, electrons will follow spiral paths which will increase the opportunity for ionization collisions between

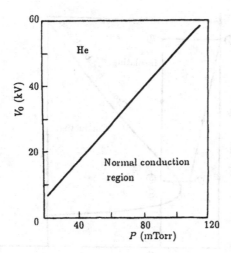

Fig.5-37 Anode voltage versus He pressure for normal ignition at a given magnetic field [72].

electrons and gas atoms. As a result, sustained discharge can be established in the tube. Application of a magnetic field, in effect, is equivalent to increasing the Pd value. Removal of the magnetic field will result in deionization of the conducting gas, hence leading to an off state. For this reason, XFT can be worked as both a close and open switch.

The operation of an XFT can be illustrated with reference to Fig.5-38 which is a plot of the voltage V_a applied to the anode or the voltage V_g to the grid versus the magnetic field B proposed by M. A. Lutz[71]. Initially, both V_a and V_g are zeroes which in the figure are denoted by (0). First V_a is increased to (1) and the tube remains insulating because B is zero. Followed by a rise of B to (2), the tube remains insulating because the magnetic field is not high enough to trap the high energy electrons and start an avalanche. Next, V_g is raised from (3) to (4), as the lower voltage present in the grid-cathode gap, this field satisfies the cross field breakdown condition and hence conduction begins. Plasma created in the grid-cathode gap is spreaded into the main gap which then begins conducting and the voltage drops to (5). Immediately thereafter, the grid voltage V_g is removed (6) and conduction continues until B is lowered to (7) at which time the tube interrupts and the voltage V_a is re-applied. At this point the on-off cycle is completed.

The data given in Table 5-7 are the demonstrated performances reported in the literature[73], for low pressure plasma opening switches (LPPOS) e.g. the tetrode

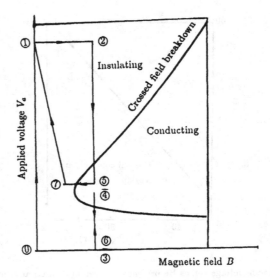

Fig.5-38 Operational sequence of events for triode XFT [71].

XFT model. These parameters are not necessarily obtainable simultaneously.

Table 5-7 Demonstrated Capabilities of LPPOS[73]

Anode Voltage at Open State	Switch Opening Time	Switch Closing Time	Repetition Rate	Total Switch Current	Life Time
120 kV	50 ns	20 ns	16 kHz	10 kA	1.5×10^9 shots

(d) Other Plasma Switches

There are a number of other switches which may fall in the category of plasma switches. They include the e-beam controlled diffuse discharge switch[28], the reflex switch[74] and the grid controlled plasma switch[75]. The first one of these switches is basically identical to the switch discussed in Subsection 5-2(b) under the sub-heading of "electron-beam triggered gaps". Further discussion of it is omitted here. In the following, we shall briefly give a short introduction for the other two switches only. For a more detailed discussion concerning the subject, one may look up the appropriate references.

The reflex switch has demonstrated the capability of a fast, high power vacuum opening switch. It can be used in conjunction with a magnetic energy storage over a wide range of currents and power levels. The basic structure of the switch consists of mainly two cathodes and one thin anode as shown in Fig.5-39. Electrons are reflected back and forth through the thin anode and positive ions are accelerated from the anode to both cathodes. Upon such reflexing of charged particles, the switch current increases drastically. This corresponds to the on state. Without reflexing of e^-, the gap behaves as a Langmuir-Child bipolar diode, this corresponds to the off state. The reflexing mode is controlled by adjusting the potential of the secondary cathode. When it is switched to the anode, the reflexing mode is turned off and the switch becomes a Langmuir-Child diode which has high impedance.

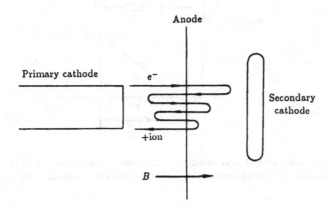

Fig.5-39 Basic structure of the reflex switch [74].

The grid controlled plasma switch is reported to be capable of working as both an open and closing switch. The basic structure and operating principle of the switch are shown in Fig.5-40. The open and closed states of the switch are controlled by the grid potential V_g. When V_g is negative with respect to the cathode potential, electrons from the plasma source are prevented from entering the power gap and only a small number of positive ions are allowed to cross the power gap. This results in a high impedance power gap and corresponds to the off state. When V_g is adjusted to the potential of the cathode, both electrons and ions from the plasma source are allowed to entering the power gap, thus conduction in the power gap enhances greatly. This corresponds to the closing state. Theoretical models have predicted capabilities of 100 kV operating voltage, 20 ns opening time and 1 μs closing time for the grid controlled plasma switches[75]. However, reports of experimental substantiation of the predicted capabilities are still lacking. Therefore, it is more appropriate to treat this switch as a new concept rather than a matured technology.

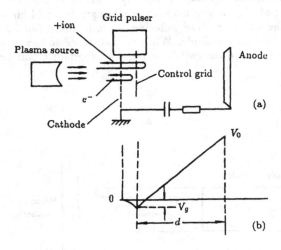

Fig.5-40 Grid-controlled plasma switch: (a) electrode configuration, load circuit, and particle flow (b) spatial variation of potential in the nonconducting state [75].

5-5 Magnetic and Repetitive Switches

(a) Magnetic Switch

The use of ferromagnetic material to produce fast pulses was first brought forth by Melville[76] in 1951. Since then significant advancements have been made in the development of the technology and the magnetic switch has now become one of the widely used devices in high power pulse applications. The basic principle governing magnetic switch is to utilize a large change in magnetic permeability exhibited by saturable magnetic material during magnetization to produce the switching effect. The basic structure and operating principle of magnetic switches are shown in Fig.5-41. It consists of mainly a ferromagnetic core surrounded with a current carrying conductor. There are various ways that the ferromagnetic core and the conductor can be arranged to form the switch. The most common types of switch configurations are the solenoid, toroid and straight conductor surrounded with ferromagnetic winding. Examples of these switch configurations are shown respectively in Fig.5-41(a), Fig.5-42(a) and (b).

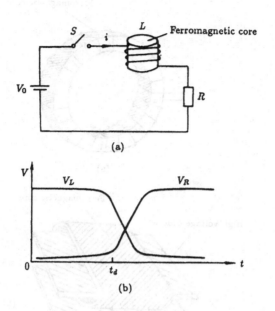

(a)

(b)

Fig.5-41 Schematic of the basic structure of magnetic switch and its operating characteristics. (a) A magnetic switch in solenoidal form with operating circuit; (b) Characteristics of voltage across switch inductor and load.

The inductance of a magnetic switch can be generally expressed as

$$L = \frac{\mu_0 \mu_r A N^2}{l} \tag{5-29}$$

In the expression μ_0 is the permeability of free space, μ_r the relative permeability of the ferromagnetic material, A the core cross section, N the number of conductor turns and l the magnetic path length. For paramagnetic and diamagnetic materials, the relative permeability μ_r is a constant. For ferromagnetic material, however, the μ_r is a function of the magnetic field H which is further dependent on current i. In Fig.5-41(a), when S closes, a time-dependent current is generated in the circuit which causes the relative permeability μ_r to change with time. However, the change of μ_r with H is not linear and its typical behavior can be seen from the B-H curve shown in Fig.5-43. For $0 < H < H_s$, since $\mu_r = B/(\mu_0 H)$, from the curve one can see that, μ_r value is relatively large and stays approximately as a constant. But, when the core reaches its saturated state, i.e. when $H \geq H_s$, μ_r value drops drastically and diminishes thereafter. From Eq.(5-29), one can see that the inductance of the switch changes correspondingly. In other

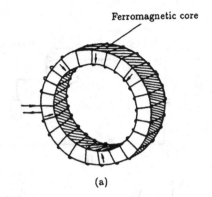

(a)

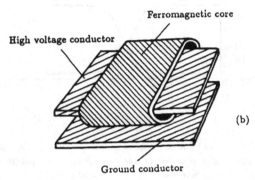

(b)

Fig.5-42 Two common configurations of magnetic switch. (a) Toroidal core with winding conductor; (b) Racetrack core with stripline conductor.

words, before the core becomes saturated, the impedance of the switch is large and most of the voltage appears across the switch. As the core becomes saturated, the switch impedance decreases rapidly and most of the voltage is switched to the load thus producing the switching effect as indicated in Fig.5-41(b). The risetime of the voltage pulse in the load may be approximately expressed as

$$t_r \approx 2.2 \frac{L_s}{R} \tag{5-30}$$

where L_s is the saturated inductance of the magnetic switch and can be further expressed by

$$L_s = \frac{\mu_0 \mu_s A N^2}{l} \tag{5-31}$$

As $\mu_s = B_s/(\mu_0 H_s)$, the risetime t_r is primarily determined by the ratio of B_s to H_s, if other parameters are given. From Fig.5-41, one can see that, after the

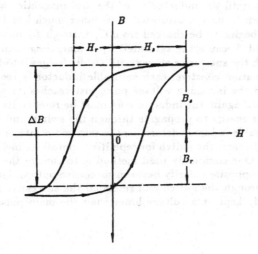

Fig.5-43 Typical *B-H* curve for a ferromagnetic material, at $H = H_s$ the material is magnetically saturated.

voltage is applied, there is a time delay before the magnetic switch closes. This is because a certain time is required for the core to be magnetized to its saturated state. The time delay t_s of the magnetic switch has been widely reported to be obtainable from the following relation[76]

$$AN\Delta B = \int_0^{t_s} V(t)dt \qquad (5\text{-}32)$$

For constant voltage $V(t) = V_0$, Eq.(5-32) reduces to

$$t_s = \frac{AN\Delta B}{V_0} = \frac{AN(B_r + B_s)}{V_0} \qquad (5\text{-}33)$$

In the expression, B_s and B_r are respectively the saturated and remanent magnetic inductions of the core as indicated in Fig.5-43. Because of the presence of B_r in the core, the magnetic switch needs to be reset (demagnetized) to its initial state before it is reused. It is usually done by applying a voltage of reverse polarity to the switch after each operation. At this point, one should note that Eqs.(5-32) and (5-33) are approximately valid only under some presumption, i.e. $\int \frac{\partial A}{\partial t}dx \equiv 0$, where A is the vector potential of the electromagnetic field. Furthermore, they are not applicable for switch configurations similar to that shown in Fig.5-42(b).

By using multiple stages of magnetic switches coupled with appropriate capacitors as shown in Fig.5-44, a slow voltage pulse can be compressed to a much faster one. The operating principle is as follows. Capacitor C_1 is first charged via the fixed inductor L_0 until the inductor L_1 of the first magnetic switch is saturated, where L_1 is chosen to have a saturated inductance much less than L_0. Once L_1 is saturated, C_2 begins to be charged from C_1 through L_1 but C_2 charges much faster than C_1 did, because the saturated L_1 is much less than L_0. The process continues through the successive stages until C_3 discharges into the load. In order to make the operation effective, each saturable inductor is required to saturate at the time when the incoming voltage pulse just reaches its peak value. If the switch is to be used again, the inductor core must be reset to its initial state. The time for the pulse energy to propagate through the switch and the time required to reset the saturable inductor determine the maximum rate of repetition. There are various ways to reset the switch for repetitive operations including the method of self-resetting. One commonly used method is to change the first capacitor in Fig.5-44(a) to an opposite polarity between successive pulses. The reverse polarity pulse cascades through the switch and resets all the successive stages. The reset pulse is commonly kept at a voltage lower than the main pulse to achieve high efficiency.

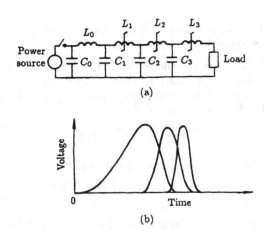

Fig.5-44 Three stages magnetic compression system [77]. (a) Basic circuit; (b) Successive voltage waveforms at each stage.

A circuit similar to that shown in Fig.5-44(a) has been quantitatively analyzed by previous authors[77]. In the analysis, all circuit components are assumed lossless and all extraneous inductances are ignored. All saturable inductors are assumed to be open circuits before saturation and the saturation takes place at peak voltages. All the saturable inductors are of toroidal geometries as indicated in Fig.5-42(a) and assuming that the area enclosed by the winding turns is simply the core cross sectional area, then the gain of the ith stage can be expressed as

$$G_i = \left(\frac{L_{i-1}}{L_i}\right)^{1/2}_{sat} = \left(\frac{V_{oli}4\Delta B_i}{\xi \pi^2 \mu_s}\right)^{1/2} \tag{5-34}$$

where $\xi = C_0 V^2/2$ is the initial energy stored in C_0. V_{oli} is the core volume of the ith inductor and can be expressed by

$$V_{oli} = \frac{\mu_{si}}{L_{si}}(N_i A_i)^2 \tag{5-35}$$

As the switch loss is proportional to the volume of the core material, it is desirable to keep the core volume minimum[78]. It can also be shown that if the core materials are identical, optimal efficiency can be achieved when all core volumes are equal. For satisfactory operation of a multistage circuit the following conditions are also required to be met.

(1) $L_{i-1} \gg L_i$ (saturated)

(2) L_{i+1}(unsaturated) $\gg L_i$ (unsaturated)

(3) saturation of inductor L_i must occur at π/ω on the charging voltage for C_i.

(4) $Ci \approx C_0$ for all i.

(5) the saturated impedance of L_i of the last stage must be much less than the load impedance.

Based on these guidelines, Birx has designed a three stage magnetic compression system of 300 J output energy[79]. The peak operating voltage of the system is 250 kV and other specifications and parameters of the system may be found in Ref[79]. For detailed information concerning switch parameters, components values and dimensions for the design of the three stage pulse compressor, one may also consult some recent publications such as Ref.[80].

In the practical design of a magnetic pulse compressor, proper selection of the core material is exceedingly important because the core characteristic is one of the main factors that determines the efficiency and capability of the system. The primary preference for the core material are large flux swings, small energy loss and low cost. The flux swing ΔB is a basic property of the material and selection of the material in this respect is rather a straightforward matter. However, selection of the material in consideration of the energy loss is not so simple, because it involves several inter-related factors. If a constant voltage V_0 is applied to the core, the energy loss due to eddy current per unit volume in the core may be expressed as[81]

$$P = \frac{\alpha \mu_s d^2 t_s V_0^2}{\sigma f_s L_s} \tag{5-36}$$

where α is a loss coefficient, σ is the electrical conductivity of the material and f_s is the stacking factor. If one defines the figure of merit as $F = \sigma f_s/(\alpha d^2 \mu_s)$ then

Eq.(5-36) can be written as $P = t_s V_0^2 / (F L_s)$ which shows that the loss is inversely proportional to the parameter F. For easy comparison, the concerned parameters for several ferromagnetic materials are listed in Table 5-8.

Table 5-8 Comparison of figures of merit for several ferromagnetic materials[81]

Material	ΔB (T)	d (μm)	α	F ($\times 10^4 \Omega$/m)
Fe-based Amorphous	2.97 (3.5)*	20	0.413	0.575
Fe-based Crystal	2.63	20	0.180	1.08
Co-based Amorphous	1.2	20	0.190	1.03
Ni-Zn Ferrite	0.72			0.524

     ⋆ after annealed.

(b) Repetitive Switch

At present, there are several types of switches which are capable of repetitive operation. However, those having high power capability, i.e. peak power in the GW range, are mostly gas spark gaps. A summary of the general performances of these switches is given in Table 5-9.

Table 5-9 Demonstrated Performances of Several Types of Switches operated under Repetitive Mode

Switch Type	Rep.rate (pps)	Voltage (kV)	Current (kA)	Insulating medium	Flow rate	Life time (shot)	Ref.
High pressure gas gap	40	530	27	Air or SF$_6$	800 (SCFH)^a	$> 10^8$	Ref.[82] This section
Tesla charged gas gap	10^3	520	3.5	N$_2$	Seavenged velocity 15 m/s	$\leq 10^8$	Ref.[84] This section
Commercially available gas gap	125	42	25	Air	200 (SCFH)	5×10^6	Ref.[83]
BLT Thyratron	10	78	19	H$_2$	Not required		Ref.[44] Section 5-2c
Cross field tube	100	70	5 (A/cm^2)	He	Not required	10^9	
Low pressure gas gap	2×10^3	20	25	H$_2$	Not required	$> 10^7$	Ref.[46] Section 5-2c
Oil gap	200	290	3	H.V. oil	1 (liter/s)		Ref.[57] This section
Magnetic switch	10^3	250	40	Ferromagnetic material	Not required	10^9	This section

     a. SCFH=Standard cubic feet per hour.

     b. Parameters are not necessarily obtainable simultaneously.

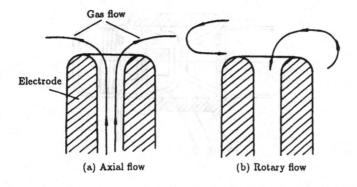

Fig.5-45 Illustration of axial and rotary gas flow patterns.

High repetitive operation of gas spark gaps requires the solutions to two basic problems, rapid recovery and removal of heat. High pressure gas spark gaps have a natural rate of recovery in the 50 to 100 Hz range. Repetitive operations beyond this range, require external means to restore the gas to its insulating state after each breakdown. The currents they conduct are usually through hot channels of partially ionized plasma whose temperature are in the order of 10^4 K. Heat created under repetitive operation from those hot channels can be tremendous and removal of heat from the active gas volume is necessary. If both the current and repetitive rate are moderate, by flushing the spark gap with sufficient gas flow is usually adequate to solve the two problems mentioned above. However, the additional measure of operations of spark gaps at either higher current or higher repetitive rate, or both, is necessary to remove the excessive heat. This is usually done by flushing the electrode body with high speed cooling flow. There are two types of gas flow patterns that are commonly employed to purge the spark gaps, axial flow and rotary flow as illustrated in Fig.5-45. The axial flow follows the shortest path from the gas source to the exhaust port, therefore high flow rate can be relatively easily achieved. In the rotary flow, the gas velocity has a large azimuthal component and the gas can effectively sweep over the entire electrode area thus preventing any effect due to uneven distribution of sparking. As both patterns have their features and advantages, the selection of gas flow pattern is largely determined by the requirements of application and availability of other facilities. In the following we shall use examples to illustrate the applications of these flow patterns in gas spark gaps.

Fig.5-46 shows the cross sectional view of a repetitive, high pressure, gas spark gap switch[82]. The switch operates at pressures of up to 10 atm with a gap length

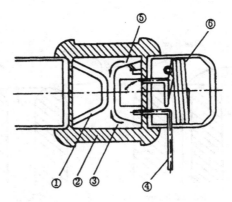

Fig.5-46 Cross sectional view of a repetitive gas spark gap [82]. (1) smaller electrode; (2) cylindrical acrylic housing; (3) gas flow pattern (rotary); (4) gas inlet; (5) gas injection port; (6) exhaust gas cooling coil.

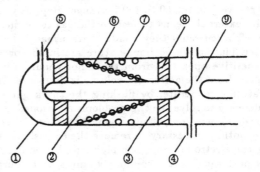

Fig.5-47 Schematic of a repetitive gas spark gap associated with a Tesla transformer [84]. (1) Grounded conductor; (2) Powered conductor; (3) H.V. oil; (4) Gas outlet; (5) Gas inlet; (6) Secondary winding of Tesla; (7) Primary winding of Tesla; (8) Insulating support; (9) Spark gap.

of 1.3 to 3.8 cm. Air or SF$_6$ was employed as the insulating medium. To achieve stable operation, a minimum gas flow rate of 800 SCFH (standard cubic feet per hour) is required. The gas is injected through 8 ports in the large electrode or through two ports in the acrylic wall (not shown in figure). The gas follow pattern is rotary type (see Fig.5-45) which can sweep and clean the housing more effectively. Exhaust gas exits through the center of the large electrode passing through a copper cooling coil located within the high voltage structure. The switch's maximum operating parameters are 700 KV for breakdown voltage, 35 kA and 30 ns for current pulse. In operation, 530 kV, 27 kA and 40 pps have been demonstrated. Typical performances are 500 kV, 13 kA and 40 pps for continuous operation over 10^5 shots.

Fig.5-47 shows an integrated unit of high power pulse facility in which both the repetitive switch and high voltage source are included in the transmission line. The high voltage source is a Tesla transformer with its primary winding attached to the inner surface of the outer conductor of the transmission line. The inner radius of the outer conductor is 30 cm. The secondary coil is wound on a hollow truncated cone and with one terminal of the winding electrically connected to the inner conductor which has an outer radius of 15 cm. The transformer winding ratio is between 10^3 to 2×10^3 and applied voltage to the primary coil is 500 V. The gap region is filled with N_2 gas at pressures of up to 12 atm. The gas is injected through one end of the inner conductor and the flow pattern is axial which facilitates high speed purging. The gas is cooled and filtered before re-injecting into the gap region. At maximum operating conditions, the active gas volume is required to be purged with a gas flow speed of 15 m/s. The switch can operate at 520 kV, and 10^3 pps, however, at this condition, the instability of the switch, i.e. the mis fire rate, becomes quite poor. Typical stable operating parameters are 520 kV 3.5 kA and 500 pps.

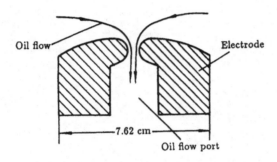

Oil flow

Electrode

7.62 cm

Oil flow port

Fig.5-48 Cross sectional view of one half of a repetitive oil spark gap [57].

The cross-sectional view of a repetitive spark gap switch using H.V. oil as the insulating medium is shown in Fig.5-48. Three different versions of similar switches have been tested in the same work[58] and one of them has gap spacing between 0.07 to 0.25 cm. The switch was purged with an oil flow of axial pattern and it was noted that the required rate of oil flow is roughly proportional to repetitive rate of operation. At repetitive rate of 200 pps, the required oil flow was between 0.63

l/s to 0.95 l/s. The upper limit to the oil flow is about 4.4 l/s which corresponds to a repetitive rate of 1250 pps. Above 4.4 l/s, cavitation occurs thus limiting the switch ability to operate at repetitive rate of 1250 pps. The demonstrated operating range of the switch are 140 to 290 kV hold-off voltage and 200 to 1250 pps repetitive rate.

References of Chapter 5

[1] I. Vitkovitsky, *High Power Switching*, (Van Nostrand Reinhold Co., New York, 1987).

[2] *Gas Discharge Closing Switches*, edited by G. Schaefer, M. Kristiansen and A. Guenther, (Plenum Press, New York, 1990).

[3] A. Von Engel, *Ionized Gases*, (Oxford, University Press, 1965).

[4] *Gaseous Electronics*, edited by M. N. Hirseh and H. J. Oskam, (Academic Press, New York, 1978).

[5] *Electrical Breakdown of Gases*, edited by J. M. Meek and J. D. Craggs, (Wiley, New York, 1978).

[6] S. T. Pai and D. T. Zhou, *Acta Physica Sinica*, 42, (1993), 1463.

[7] J. S. Townsend, *The Theory of Ionization of Gases by Collission*, (Constable, London, 1910).

[8] J. M. Meek, *Phys. Rev.* 57, (1940), 722.

[9] H. Raether, *Electron Avalanches and Breakdown in Gases*, (Butterworths, London, 1964).

[10] L. B. Loeb, *Fundamental Processes of Electrical Discharges in Gases*, (Wiley, New York, 1939).

[11] V. V. Kremnev and G. A. Mesyats, *Zh Prikl. Mekh. Tekh. Fiz.* 1, (1971), 40.

[12] E. E. Kunhardt and W. W. Byszewski, *Phys.Rev.* A21, (1980), 2069.

[13] S. T. Pai, *J. Appl. Phys.* 71, (1992), 5820.

[14] S. T. Pai, *J. Tsinghua Univ.* (in Chinese) 33, (1993), 11.

[15] T. H. Martin, *5th IEEE Pulsed Power Conf.*, (1985), 74.

[16] F. E. Peterkin and P. F. Willaims, *7th IEEE Pulsed Power Conf.*, (1989), 559.

[17] R. E. Wootton, *5th IEEE Pulsed Power Conf.*, (1985), 258.

[18] J. Gruber and G. Mueller, *Max Planck Inst. Plasma Phys. Garching Report*, IPP 4/27, (1965).

[19] W. Wright and R. Hitchcock, *IEEE 16th Power Modulator Symp.*, (1984), 38.

[20] Wang Jia, *J. Tsinghua Univ.* (in Chinese) 22, 3, (1982), 59.

[21] A. I. Pavlovskii et al., *Pribory i Tekhnika Eksperimenta*, 2, (1970), 122.

[22] S. Mercer et al., *Proc. Int. Conf. Energy Storage Compression and Switching*, Torino, (1974), 458.

[23] V. G. Emelyanov et al., *Izvestiya Vysshikh Uchebnykh Zavedenii Fizika*, 17, (1974), 136.

[24] A. S. Elchaninov et al., *Sov. Phys. Tech. Phys.* 20, (1975), 51.

[25] Y. Tzeng et al., *IEEE Trans. on Plasma Sci.* PS-10, (1982), 234.

[26] Y. Tzeng et al., *3rd IEEE Pulsed Power Conf.* (1981), 231.

[27] R. J. Commisso et al., *Rev. Sci. Inst.* 55, (1984), 1834.

[28] G. Schaefer and K. H. Schoenbach, *IEEE Trans. on Plasma Sci.* PS-14, (1986), 561.

[29] A. H. Guenther and J. R. Bettis, *Proc. IEEE* 59, (1971), 689.

[30] A. H. Guenther and J. R. Bettis, *J. Phys.* D.11, (1978), 1577.

[31] A. H. Guenther and J. R. Bettis, *Int. Conf. on Phenomena in Ionized Gases* (Invited Papers), (Swansea, 1987).

[32] J. R. Woodworth et al., *IEEE Trans. Plasma Sci*, PS-10, (1982), 257.

[33] J. R. Woodworth et al., *J. Appl. Phys.* 56, (1984), 1382.

[34] B. N. Turman et al., *8th IEEE Pulsed Power Conf.*, (1991), 319.

[35] R. S. Taylor and K. E. Leopold, *Rev. Sci. Instrum*, 55, (1984), 52.

[36] A. Endoh and S. Watanabe, *J. Appl. Phys.* 59, (1986), 3561.

[37] H. J. Doucet et al., *Laser and Particle Beams* 5, (1987), 451.

[38] Wang Jia and S. T. Pai, *6th Int. Symp. on High Voltage Eng.*, (New Orleans, 1989), p.22.25.

[39] S. T. Pai and J. P. Marton, *IEEE Trans on Electrical Insulation*, **EI-20**, (1985), 93.

[40] S. T. Pai and J. P. Marton, *15th IEEE Power Modulator Symp.*, (1982), 153.

[41] L. M. Earley and G. L. Scott, *8th IEEE Pulsed Power Conf.* (1991), 340.

[42] Win Shongtao, *Construction and Evaluation of a Multi-stage, Multichannel High Voltage Switch* (in Chinese), M.Sc.Thesis, (Tsinghua University, Beijing, 1988).

[43] J. M. Wilson and G. L. Donovan, *6th IEEE Pulsed Power Conf.*, (1987), 361.

[44] G. Kirkman-Amemiya et al., *8th IEEE Pulsed Power Conf.*, (1991), 482.

[45] J. Christiansen and C. Schultheiss, *Z. Physik*, A **290**, (1979), 35.

[46] K. Frank et al, *8th IEEE Pulsed Power conf.*, (1991), 472.

[47] E. Boggasch and H. Riege, *Proc. XVIIth Int. Conf. on Phenom. in Ionized Gases*, (Budapest, 1985).

[48] G. F. Kirkman and M. A. Gundersen, *Appl. Phys.Lett.* **49**, (1986), 494.

[49] D. L. Lorce et al., *IEEE 19th Power Modulator Symp.*, (1990).

[50] M. A. Lutz, *IEEE Trans Plasma Sci.*, PS-5, (1977), 273.

[51] A. P. Alkhimov et al., *Sov. Phys. Doklady*, **15**, (1971), 959.

[52] J. C. Martin, *Pulsed Electric Power Circuit and Electromagnetic System Design Notes PEP 4-1*, **1**, Note 4, (1973).

[53] G. Herbert, AWRE Note SSWA/HGH/6610/104, (1966).

[54] D. L. Johnson et al., *IEEE Trans Plasma Sci.*, PS-8, (1980), 204.

[55] L. S. Levine and I. M. Vitkovitsky, *IEEE Trans on Nucl. Sci.* NS-18, (1971), 225.

[56] D. M. Weidenheimer et al., *9th Int.Conf. on High Power Particle Beams*, **1**, (Washington D.C., 1992), 640.

[57] P. D'A.Champney et al., *8th IEEE Pulsed Power Conf.*, (1991), 863.

[58] R. D. Ford and M. P. Young, *Proc. 8th Symp. on Fusion Technology*, (Luxemburg, 1974), 377.

[59] T. E. James, Culham Laboratory (Abington U.K.) Report CLM-23, (1973).

[60] R. A. Richardson et al., *8th IEEE Pulsed Power Conf.*, (1991), 187.

[61] D. M. Giorgi et al., *8th IEEE Pulsed Power Conf.*, (1991), 122.

[62] M. K. Browder and W. C. Nunnally, *8th IEEE Pulsed Power Conf.*, (1991), 1024.

[63] M. K. Browder and W. C. Nunnally, *7th IEEE Pulsed Power Conf.*, (1989), 433.

[64] F. J. Zutavern et al., *7th IEEE Pulsed Power Conf.*, (1989), 412.

[65] G. M. Loubriel et al., *8th IEEE Pulsed Power Conf.*, (1991), 33.

[66] P. F. Ottinger et al., *J. Appl. Phys.* **56**, (1984), 774.

[67] B. V. Weber et al., *Laser and Particle Beams* **5**, (1987), 536.

[68] C. W. Mendel et al., *Rev. Sci. Instrum.* **51**, (1980), 1641.

[69] F. Vernneri et al., *4th IEEE Pulsed Power Conf.*, (1983), 350.

[70] J. C. Cochrane et al., *8th IEEE Pulsed Power Conf.*, (1991), 618.

[71] M. A. Lutz, *IEEE Trans Plasma Sci.* PS-5, (1977), 273.

[72] R. J. Harvey and M. A. Lutz, *IEEE Trans Plasma Sci.* PS-4, (1976), 210.

[73] *Opening Switches*, Edited by A. Guenther, M. Kristiansen and T. Martin, (Plenum Press, New York, 1987).

[74] J. Creedon et al., *J. Appl. Phys.* **57**, (1985), 1582.

[75] S. Humphries, JR et al., *IEEE Trans Plasma Sci*, **PS-13**, (1985), 177.

[76] W. C. Nunnally, *3rd IEEE Pulsed Power Conf.*, (1981), 210.

[77] D. L. Birx et al., *3rd IEEE Pulsed Power Conf.*, (1981), 262.

[78] D. M. Barrett, *8th IEEE Pulsed Power Conf.*, (1991), 735.

[79] D. L. Birx, *Basic Principles Governing the Design of Magnetic Switches*, UCID 18831, (1980).

[80] H. C. Kirbie et al., *7th IEEE Pulsed Power Conf.*, (1989), 171.

[81] H. Seki et al., *8th IEEEPulsed Power Conf.*, (1991), 727.

[82] M. T. Buttram and G. J. Rohwein, *IEEE 13th Pulse Power Modulator Symp.*, (1978), 303.

[83] Maxwell Lab. Inc. Spark Gap Switch 40264 R Data Sheet.

[84] N. M. Bykov et al., *Pribory i Tekhnika Eksperimenta*, No.2 (1991), 38.

CHAPTER 6
DIODE AND PARTICLE BEAMS

6-1 Electron Emission Processes

(a) Field Emission of Cold Surface

In an e-beam diode, the basic mechanism that is responsible for generation of electrons is the field emission process. The theory of field emission from a cold metal surface was developed originally in 1928 by Fowler and Nordheim on the basis of quantum mechanical tunneling of conduction electrons through the potential barrier at the metal surface[1]. The current density produced by field emission process is given by

$$J = AE^2 \exp(-B/E) \tag{6-1}$$

In the expression E is the electric field at the metal surface, A in (Amp/Volt^2) and B in (V/m) are constants and related to the work function Φ and Fermi energy ϵ_f (both are expressed in electron volts) of the metal by the expressions

$$A = \frac{6.2 \times 10^{-3}}{\Phi + \epsilon_f} \left(\frac{\epsilon_f}{\Phi} \right)^{1/2} \tag{6-2}$$

and

$$B = 6.8 \times 10^9 \Phi^{1.5} \tag{6-3}$$

For graphite, which is frequently used as cathode material, its Φ and ϵ_f values are approximately 4 and 12eV, respectively.

According to the theory, upon application of the external field at the metal surface, the width of the potential barrier will be reduced which enables the electrons to tunnel through it more easily. Once the electron tunnels through the potential barrier it becomes a free electron. The estimated electric field required for the electrons to tunnel through a typical barrier of 1 nm width is about $3 \times 10^7 \text{V/cm}$. However, in many practical cases, field emission of electrons from cathode surface occurs at gap fields that are far lower than the required threshold field of $3 \times 10^7 \text{V/m}$. For example, pre-breakdown current was frequently observed at field strengths of about 10^5V/cm in vacuum gaps. The main reason causing such a difference is that the electrode surface is far from planar and there exists various microscopic protrusions whose tip radii are usually much smaller than the base radii as indicated in Fig.6-1. Upon application of high voltage to the gap, field lines would converge on the tips of these microscopic protrusions and thus the local fields at those tips are significantly enhanced. The local field E_m at the tip can be expressed as

$$E_m = fE \tag{6-4}$$

where E is the macroscopic gap field, f is the field-enhancement factor. For an ideal microprotrusion as that shown in Fig.6-1, the field-enhancement factor is given by[2]

$$f = \frac{(\lambda^2 - 1)^{1.5}}{\lambda \ln[\lambda + (\lambda^2 - 1)^{1/2}] - (\lambda^2 - 1)^{1/2}} \qquad (6\text{-}5)$$

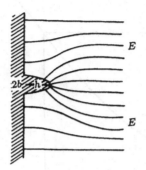

Fig.6-1 Illustration of field lines converge on an ideal microprotrusion [2].

where $\lambda = h/b$. For needle-like geometries, whose $\lambda \gg 10$, the above expression is reduced approximately to

$$f \approx \frac{\lambda^2}{\ln \lambda - 0.3} \qquad (6\text{-}6)$$

A plot from Eq.(6-5) is shown in Fig.6-2. From the figure, one can see that at about $\lambda = 20$, the local field E_m at the tip of a microprotrusion is about two orders of magnitude greater than the macroscopic gap field E. This explains the fact that the pre-breakdown current was frequently observed at applied field strength of as low as 10^5V/cm. This process, however, is important only in the initial phase when the external field is turned on. Once the field emission process takes place, the microprotrusion will be rapidly evaporated due to Joule heating, then a different emission process termed as an explosive electron emission will become predominant which will be discussed in the next section.

(b) Explosive Electron Emission

As the local field at the tip of a microprotrusion (or whisker) approaches 10^7V/cm, current due to the field emission of electrons begins to flow through the whisker thus the temperature of the whisker increases. The maximum temperature of a whisker, having similar geometries as that shown in Fig.6-1, can be approximately expressed by[3]

$$T_m = \frac{1}{2}\left(\frac{J^2 \eta}{K}\right) h^2 \qquad (6\text{-}7)$$

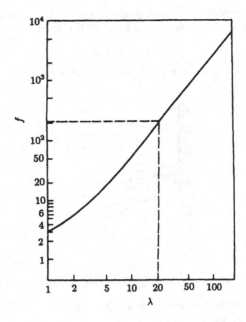

Fig.6-2 A plot of f versus λ from Eq.(6-5).

where η and K are respectively the electrical resistivity and thermal conductivity of the whisker material, h is the height of the whisker as indicated in Fig.6-1. For the case of a tungsten cathode whose melting point is about 3400°C, one may take $\eta = 5 \times 10^{-5}\,\Omega\cdot\mathrm{cm}$ and $K = 0.25$ cal/sec.cm°C. A typical whisker has the height in the range 10^{-4}–10^{-3}cm. By putting these values into Eq.(6-7), one can find that a current density of the order of 10^7–$10^8\,\mathrm{A/cm^2}$ is sufficient to melt the whisker into vapor. The material in the vapor state is quickly ionized to form a local plasma flares in the vicinity of the melted whiskers. These individual flares rapidly expand and merge into a plasma cloud that spreads over the entire cathode surface. This cathode plasma may be considered as an electron emitter having zero work function. Therefore electron current can readily be extracted from this plasma and it is limited only by the space charge cloud in the anode cathode gap. For a planar gap, the current density is limited by the Child-Langmuir formula, namely

$$J = \frac{4\epsilon_0}{9}\left(\frac{2e}{m_0}\right)^{1/2}\frac{V^{3/2}}{d^2} \qquad (6\text{-}8)$$

where ϵ_0 is the permittivity of free space, e and m_0 are respectively the charge and rest mass of the electron, V the applied voltage and d the gap spacing.

The expression given in Eq.(6-7) is a steady state equation which describes the relationship between the maximum temperature of the whisker and the current density going through it at equilibrium condition. It did not specify how much time is required for the whisker to reach such temperatures after the current begins to flow in it. In practice, there is always a time interval between the beginning of current flow and the moment at which the whisker explodes due to Joule heating. This time interval is called point explosion delay time which is denoted by t_d and expressed approximately by the relation[4]

$$\int_0^{t_d} J^2(t)dt = 0.55(\pi^4 \rho c)/\eta \qquad (6-9)$$

If the pre-explosion emission current $J(t)$=constant, then Eq.(6-9) becomes

$$t_d = 0.55 \left(\frac{\pi^4 \rho c}{\eta J^2} \right) \qquad (6-10)$$

where ρ, c, and η are the density, the specific heat and the electrical resistivity of the cathode material, respectively. This point explosion delay time is important for the situations where electron beam pulse of fast risetime is required such as in the case of excimer laser pumped with electron beam pulses. Because point explosion delay time is closely related to the breakdown delay time of a vacuum gap which is usually employed to generate electron beam.

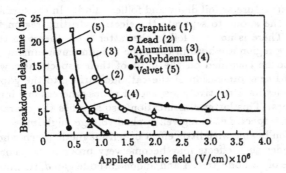

Fig.6-3 Experimental breakdown delay time as a function of applied field and electrode materials [5].

From Eq.(6-10), one can see that the delay time t_d is inversely proportional to the electrical resistivity of the cathode material, therefore with cathodes made of high resistivity material, the breakdown delay time of a vacuum diode can be significantly reduced. Fig.6-3 shows the results of breakdown delay times for several materials that may be employed as cathode material. One can see from the figure that the insulating material Velvet has the best performance in this respect.

Recent study[5] has further demonstrated the suitability of Velvet as a cathode material. In the study electron beam pulse of 3 ns duration and 7 kA/cm^2 was obtained from a Velvet cathode.

6-2 Electron Beam Diode

(a) General Remarks

One of the major goals in developing high power pulse technology is to produce high power particle beams, such as electron beams in the TW range. To convert high power electromagnetic pulses to electron beams (e-beams) requires the use of e-beam diode. The basic structure of the e-beam diode consists of a cathode and an anode (with or without a thin foil) which are separated by a vacuum gap. The vacuum requirement is moderate and is usually in the range of 10^{-5} to 10^{-3} torr. Upon application of high enough voltage to the electrodes, field emission of electrons from the cathode will take place which produces a plasma near the cathode surface. This plasma in turn serves as an electron source and leads to the production of a more intense stream of electrons moving toward the anode. At high energies, the electrons can readily penetrate a thin anode foil and continue on to form an electron beam with a beam current up to several mega-amperes and in the tera-watt power range.

Basing on applications and requirements, many types of e-beam diode have been developed in the past. Electron beam diodes may be classified on the basis of their diode structure as foil diode and foilless diode. In the former, a thin metal foil is used as the anode to separate the electron drifting region from the active diode region. There is no such foil in the latter type, therefore the diode region and the drifting region are both in the same vacuum environment. Electron beam diodes can also be classified on the basis of their aspect ratio which is defined as the ratio of linear dimension of the cathode surface to the gap spacing. High impedance diodes have relatively low aspect ratios whereas low impedance diodes have higher ones. This latter classification also signifies the characteristics of the diodes in other respect, i.e. at a given applied voltage, high impedance diodes may produce electron beams with little effect of self-magnetic pinch whereas in the low impedance diodes such effects may become very pronounced. For nonrelativistic beams, a diode of cathode area A and anode-cathode gap d, its impedance is given approximately by[6]

$$Z = \frac{427}{\sqrt{V}} \left(\frac{d^2}{A} \right) \quad (\Omega) \tag{6-11}$$

where V is the gap voltage in (MV). From this expression, one can see that, at a given voltage V, the diode impedance Z is inversely proportional to the aspect ratio. For a circular cathode, it is R/d where R is the radius of the circular cathode. In the following sections, we shall discuss the basic structures and characteristics of each type of those e-beam diodes respectively.

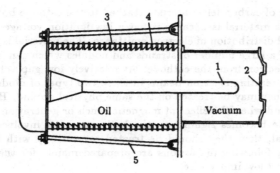

Oil   Vacuum

Fig.6-4 High impedance e-beam diode with foil anode. 1, Cathode; 2, Foil anode;
  3, Insulating ring; 4, Grading ring; 5, Insulator tied rod [13].

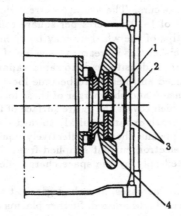

Fig.6-5 Low impedance e-beam diode with foil anode. 1, Cathode; 2, Carbon felt;
  3, Anode foils; 4, Shielding cover [14].

(b) Diode with Anode Foil

Diodes with anode foil are generally classified on the basis of their aspect
ratios, i.e. high impedance and low impedance diodes. In applications, high
impedance diodes are generally employed to produce e-beams of relatively high
energy and low current whereas low impedance diodes are more frequently used
for the generation of large current and highly magnetically pinched e-beams. The
basic structural difference between the two types of diode is of the aspect ratio,
i.e. the ratio of cathode radius to gap spacing. In the high impedance diodes it
is much smaller than that in the low impedance diodes. Fig.6-4 and Fig.6-5 show
respectively the basic structures of the two types of e-beam diode. In Fig.6-5,

there is a layer of carbon felt material attached to the cathode body as indicated. The use of this material is because it has a low ignition voltage and reasonably uniform spatial distribution of electron emittance. The material is fabric in nature and consists of a large number of strands and nodules which can greatly facilitate emissions of electron from the cathode. At relatively low gap voltages, the basic processes in generating electron beams in these two types of diode are essentially identical and they may be treated on the same physical ground. However, at high gap voltages and when the effect of magnetic pinch or generation of positive ions from the anode becomes significant, the basic processes involved are no longer totally identical, then they should be treated respectively with the appropriate approach. The following discussions are prepared mainly for the type of diodes having relatively low impedance.

As mentioned before, a wide variety of whiskers and micro-imperfections, whether natural ones or artificially produced, are present on the cathode surface. When high voltage pulse is applied across the anode-cathode gap, field emission of electrons from whiskers occurs. The average current density at this point is low, typically is in the range of 1–10 amperes per square centimeter, though the local current density at the tips of the whiskers may be as high as 10^7 A/cm^2. As the enhanced field at the whisker tips increases to a value of about 10^8 V/cm, resistive heating due to large current density leads to rapid vaporization of those whiskers. The vapor is easily ionized and heated to become local plasma bursts and flares. Subsequent rapid expansion and the merger of these local plasma flares quickly to form a plasma sheath which covers the entire surface of the cathode. Observations show that the plasma sheath expands quickly toward the anode with a velocity of 1–4 cm/μs. Since this plasma sheath is effectively equivalent to a metal surface of zero work function, electron current supplied from the cathode plasma can be quite large and is limited only by the space charge effect of the Child-Langmuir law.

As the beam current increases, absorbed gases at the anode as well as some anode material are released and ionized, further plasma originated from the anode is formed. This anode plasma moves toward the cathode with a velocity usually less than that of the cathode plasma. Resulting from the motions of both cathode and anode plasmas toward each other, the effective gap spacing d decreases with time and is approximately expressed as

$$d = d_0 - v_p t \tag{6-12}$$

where d_0 is the original gap spacing and v_p is the plasma velocity. As a consequence, the expressions given in Eq.(6-8) and Eq.(6-11) for the current density and diode impedance also need to be modified as

$$J = \frac{4\epsilon_0}{9} \left(\frac{2e}{m_0}\right)^{1/2} \frac{V^{3/2}}{(d_0 - v_p t)^2} \tag{6-13}$$

and

$$Z = \frac{427 \times 10^3}{A\sqrt{V}} (d_0 - v_p t)^2 \tag{6-14}$$

At a moderate gap voltage, if d_0 is large and v_p is small, from Eq.(6-13) the current density may be low. Since the self-magnetic field in this case is low, the pinch effect is unimportant and the electrons stream across the gap reaching the anode with practically a zero angle of incidence as indicated with dashed lines in Fig.6-6. However, if d_0 is small or V_p is large or both, the magnetic force can no longer be ignored and the electron trajectory will become curved toward the diode axis. When the centripetal force at the beam edge equals to the magnetic force i.e.

$$evB_\theta = \frac{m_0 v^2}{d_0} \qquad (6\text{-}15)$$

the electrons initiated at the cathode edge (outermost electrons) will follow a circular trajectory of radius d_0 and impinge on the anode tangentially as shown in Fig.6-6. In Eq.(6-15), B_θ is the azimuthal self-magnetic field induced by current density J.

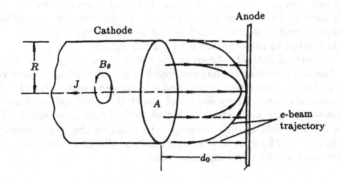

Fig.6-6 An illustration of the e-beam trajectories with and without self-magnetic pinch effect in a cylindrical diode.

When taking relativistic effect into consideration Eq.(6-15) becomes

$$\beta\gamma = \frac{ed_0}{m_0 c} B_\theta \qquad (6\text{-}16)$$

where $\beta = v/c$, relativistic factor $\gamma = 1/(1 - \beta^2)^{1/2} = 1 + eV/(m_0 c^2)$ and c is the speed of light. By substituting $B_\theta = \mu_0 I/(2\pi R)$ into Eq.(6-16), one obtains the total beam current I at this stage as

$$I = \frac{2\pi m_0 c}{\mu_0 e} \beta\gamma \frac{R}{d_0} \qquad (6\text{-}17(a))$$

or

$$I_c \approx 8500 \times \beta\gamma \frac{R}{d_0} \qquad (6\text{-}17(b))$$

This current is usually denoted by I_c and is termed as the critical current. When the diode current exceeds I_c, the effect of self-magnetic pinch will become important and the e-beam is then self-focusing as indicated in Fig.6-6. The expression given in Eq.(6-17) has been shown to be approximately valid provided one takes the time-dependent factor, namely replacing d_0 by $d = d_0 - v_p t$, into consideration. This expression is valid for diodes having cylindrical cathode. For diodes with rectangular cathode of dimensions $W \times l$, the corresponding expressions are[7]

$$\left. \begin{aligned} I_c' &= 2700\beta\gamma\frac{l}{d_0} \quad \text{(for long edge)} \\ I_c'' &= \frac{8500\beta\gamma}{1+\ln(2l/W)}\frac{l}{d_0} \quad \text{(for short edge)} \end{aligned} \right\} \tag{6-18}$$

It should be noted that, when the diode current exceeds the value given in Eq.(6-17) or Eq.(6-18), self-magnetic pinch of the e-beam always occurs, however the pinch may not always be symmetric about the diode axis or always collapses to a very small radius. Production of the e-beam on the axis may be achieved when a thin dielectric rod cathode is used as the beam cathode. However, use of this approach is limited to high impedance ($\geq 20\Omega$) diodes.

If the cylindrical cathode shown in Fig.6-6 is replaced by an annular cathode of the configuration similar to that employed in Ref.[8] and the aspect ratio of the diode is kept large, a more tightly pinched e-beam can be achieved (Fig.6-7). Such pinched beams are generally symmetrical about the diode axis and are important for applications where accurate focusing of the e-beam is required. There have been several theoretical models developed to describe the characteristics of such pinched beams[8,9]. The general conclusion is that the diode current of this nature can be described by the following expression

$$I = 8500gr\ln[\gamma + (\gamma^2 - 1)^{1/2}] \tag{6-19}$$

where g is a geometrical factor depending on the details of the diode geometry. For the conical diodes with planar anode shown in Fig.6-7 $g = R/d_0$ in the parapotential flow model[8], and $g = R/[d(\gamma)^{1/2}]$ in the focused flow model[9]. The diode impedance in this case is given by

$$Z = \frac{V}{8500gr\ln[\gamma + (\gamma^2 - 1)^{1/2}]} \tag{6-20}$$

According to the parapotential flow model, electrons move along equipotential surfaces in a region extending from the vicinity of the cathode to the vicinity of the anode. In order to satisfy this condition, the resultant force $e(E + v \times B)$ acting on the electrons was assumed to be zero, i.e. the electrostatic force was just balanced by the Lorentz force. The model prediction has been found to be in fair agreement with measurements of diode impedance. However, it is unable to give explainations as to how electrons get on and off equipotential surfaces when they

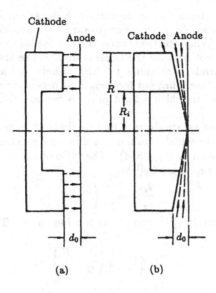

Fig.6-7 Pinched e-beam diode with (a) flat and (b) conical front surfaces. The hollow well of radius Ri helps reduce the effects of plasma motion [8].

are very near the electrodes. Nor can it account for the effects of ion flow, which have been found to be very important for large aspect ratio diodes. The focused-flow model is applicable only for solid- cathode configurations and for a steady state current flow following the pinch completion, therefore it is also incapable of adequately explaining the physical mechanism that is responsible for the initial process of the magnetic pinch. As a result, a newer model which describes the time-depended evolution of the flow in terms of four distinct phases was developed by SNL researchers[10]. The basic features of the new model may be summarized as follows.

Phase 1, Child-Langmuir electron flow at low voltage. This phase corresponds to the early stage and the gap voltage is relatively low. The current is contributed from electrons only and the flow is orthogonal to the equipotentials as shown in Fig.6-8(a). The current density can be approximately described by the modified Child-Langmuir formula given in Eq.(6-13).

Phase 2, Electron flow at high voltage with weak pinch. As the gap voltage increases, at a point the diode current exceeds the critical current I_c given by Eq.(6-17) and the self-magnetic field bends the outer electron trajectories to form a weak pinch as shown in Fig.6-8(b). At this time, the diode current is expressed by Eq.(6-19).

Phase 3, Electron flow begins to collapse due to ion flow from anode plasma.

As energetic electrons strike the anode surface, a dense anode plasma is formed. Ions emitted from the anode plasma flow toward the cathode and neutralize the electron space charge. When electrons enter the anode plasma at grazing incidence, the electric force may not balance the Lorentz force hence they are reflected back toward the cathode while continuing to drift radially inward until a region is reached where anode plasma has not formed. This process is repeated until finally the condition is reached in which a tight pinch is formed. This is shown in Fig.6-8(c).

Phase 4, Quasi-steady state electron and ion flows. A quasi steady state bipolar flow is finally reached in which laminar ion flow and pinched electron flow form the total current as shown in Fig.6-8(d) . The ratio of the ion to electron currents can be approximately expressed by[11]

$$\frac{I_i}{I_e} \geq \frac{1}{2} \frac{R}{d_0} \left(\frac{\beta'}{\beta} \right) \tag{6-21}$$

where $\beta' = [2(\gamma - 1)m_0/m_i]^{1/2}$ and m_i is the ion mass. The electron and ion current are given by

$$I_e = I_p \left\{ 1 + \frac{1}{2} \left(\frac{\beta'}{\beta} \right) \frac{R}{d_0} \right\}^{-1} \tag{6-22}$$

and

$$I_i = I_p \left\{ 1 + 2 \left(\frac{\beta}{\beta'} \right) \frac{d_0}{R} \right\}^{-1} \tag{6-23}$$

where I_p is the saturated parapotential current given by Eq.(6-19).

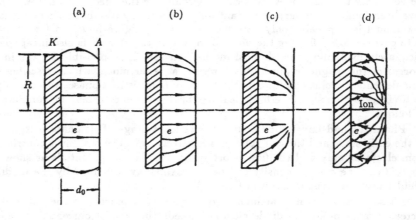

Fig.6-8 Evolution of charge flow in a tightly pinched e-beam diode. (a) Child-Langmuir electron flow at low voltage; (b) Electron flow at high voltage with weak pinch; (c) Anode plasma induced collapsing electron ring; (d) Quasi-steady state electron and ion flows.

The above description is valid only for diodes having large aspect ratio, i.e. R/d_0 is significantly greater than unity for diodes of small aspect ratio, such as the one shown in Fig.6-4, one can use the formulas given in Eq.(6-13) and Eq.(6-14) for diodes with expanding plasmas or

$$J_e = 1.86J_0 \tag{6-24}$$

and

$$J_i = 1.86J_0(m_0/M)^{1/2} \tag{6-25}$$

for the cases bipolar space-charged-limited flow is significant. In these expressions J_e, J_i and J_0 are respectively the current densities of electron, positive ion, the Child-Langmuir expression given by Eq.(6-8) and M is the mass of the ions. If electron flow to the anode wall is magnetically cutoff, then the total current I flowing in the diode is given by[8]

$$I == \frac{8500\gamma_m}{\ln(R'/R)} \left\{ \ln[\gamma_m + (\gamma_m^2 - 1)^{1/2}] + \frac{\gamma - \gamma_m}{(\gamma_m^2 - 1)^{1/2}} \right\} \tag{6-26}$$

where γ_m is the value of γ at the maximum extent of the flow pattern, R' is the radius of the diode vacuum chamber. However, the use of any formula given in this section (Section 6-2) is subject to possible gross errors, because all these formulas were derived on the basis of steady state solutions of some fundamental equations which strictly speaking have no provision to deal with the dynamic behaviors of electrons and ions in the diode.

The following are two examples to show the operating parameters of two practical diodes of similar configurations as that shown in Fig.6-4 and Fig.6-5, respectively. Table 6-1 lists the various parameters of a high impedance diode built by the China Academy of Engineering Physics (CAEP)[13] having a similar configuration as that shown in Fig.6-4. The experimental results obtained by the same authors from the diode are given in Table 6-2. The critical current given in the table was obtained from Eq.(6-17) using the parameters $\gamma = 12.5$ and $\beta = 0.9968$. As the observed diode current is considerably greater than the critical current, indicating that appreciable magnetic pinch was present in the process. This pinch is essential to prevent electrons emitted from the cathode shank to reaching the diode chamber wall.

Table 6-1 Parameters of a high impedance diode of similar configuration shown in Fig.6-4 [13]

Cathode		Anode		Gap Spacing	Diode Chamber Diameter	Vacuum Condition
Radius	Shank	Material	Thickness			
8.45cm	330cm	Titanium	0.5–0.8mm	18cm	114cm	$3–6 \times 10^{-3}$ Pa

Table 6-2 Experimental results obtained from the diode with its parameters shown in Table 6-1

Diode Voltage (MV)	Diode Current (kA)	Diode Impedance (Ω)	Critical Current (kA)
5.9	77.6	76	49.7

The second example is for a low impedance diode of similar configuration shown in Fig.6-5. This diode was built by the Institute of Atomic Energy of China[14] and has a 36×6 cm^2 carbon felt cathode and a 40×8 cm^2 graphite anode with a 25μm Ti foil. The observed and calculated results obtained by the same authors from this diode are given in Table 6-3. The diode current, impedance, critical currents I_c' and I_c'' shown in the table were calculated respectively from Eq.(6-8), Eq.(6-11) and Eqs.(6-18) using the parameters $\gamma = 2.176$ and $\beta = 0.888$. From Table 6-3, one can see that when the gap spacing is less than 1.4 cm, the effect of self-magnetic pinch begins to be appreciable as at $d_0 = 1.2$ cm the observed current is already greater than the critical current I_c''.

Table 6-3 Results obtained from a low impedance diode of similar configuration shown in Fig.6-5 [14]

Anode-Cathode Spacing (cm)	Diode Voltage (kV)	Diode Current		Diode Impedance		Critical Current I_c' (kA)	Critical Current I_c'' (kA)
		Observed (kA)	Cal. (kA)	Observed (Ω)	Cal. (Ω)		
1.2	597	147	162	4.1	3.7	156	141
1.4	593	119	119	5.0	5.0	134	121
1.6	592	91	92	6.5	6.6	117	106

(c) Foilless Diode

One of the disadvantages associated with the e-beam diode having anode foil is anode vaporization and beam heating due to anode foil scattering of electrons. Diodes without anode foil naturally do not have such a disadvantage. Besides, foilless diodes generate annular e-beams which can supply greater space charge limited current than that of a solid beam carrying the same number of charged particles. The disadvantage of foilless diode is that it requires externally applied magnetic field to control the expansion and transport of the e-beam whereas for the diodes with anode foil, the external magnetic field is not necessary. Because there is no anode foil, the e-beam drifting region is not separated from the diode active region and both of them are required to be under the same vacuum condition.

There have been a number of foilless diodes in different geometries developed in the past. Two typical designs of foilless diode are shown in Fig.6-9 and Fig.6-10. The first one is a planar foilless diode which is designed to inject electrons approximately parallel to the applied uniform field. The electron beam so produced is usually in an annular form and can propagate in a vacuum tube under the guide field. The second configuration is essentially a magnetron gun in which the electrons are injected orthogonally to the magnetic field. Because the magnetic field is strong, the electrons do not cross the diode gap, but rather drift into the guide region following the curved magnetic field lines as indicated in Fig.6-10. In order to facilitate emission of electrons, there is usually a roughened area on the cathode to supply the electrons. The diode current for a solid beam is given by[12]

$$I = \frac{17000(\gamma^{2/3} - 1)^{3/2}}{1 + 2\ln(R/r_2)} \tag{6-27}$$

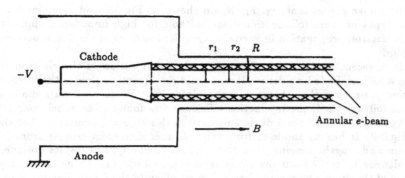

Fig.6-9 Foilless diode in which emission is parallel to the applied magnetic field.

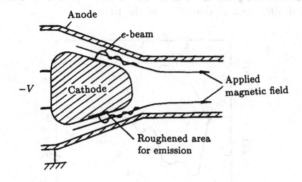

Fig.6-10 Foilless diode in which emission occurs orthogonal to the applied magnetic field.

While for a hollow beam of thickness $a \ll r_1$, Eq.(6-27) becomes

$$I = \frac{17000(\gamma^{2/3} - 1)^{3/2}}{a/r_1 + 2\ln(R/r_2)} \qquad (6\text{-}28)$$

If $(r_2 - r_1) \ll (R - r_2)$, the diode impedance can be approximated by

$$Z \approx \frac{(R - r_2)}{4\pi c r_b \epsilon_0} \ln\left[\frac{8(R - r_2)}{a}\right] \qquad (6\text{-}29)$$

and the total diode current is then

$$I = \frac{17000 r_b (\gamma^{2/3} - 1)^{3/2}}{(R - r_2)} \left\{\ln\left[\frac{8(R - r_2)}{a}\right]\right\}^{-1} \qquad (6\text{-}30)$$

In the above expressions, r_1, r_2, R are shown in Fig.6-9 and $r_b = [r_1 + r_2]/2$. Both types of these foilless diodes are suitable for high impedance applications when electron propagation in vacuum is required such as in the linear accelerator system.

In a recent study it was demonstrated that injection of the electron beam from a foilless diode into dense plasma could significantly enhance the interaction efficiency between the e-beam and the plasma[15]. When a vacuum diode with anode foil was used, the interaction efficiency was found to be much lower. The configuration of the foilless diode employed in this case was similar to that shown in Fig.6-9. It has an anode drifting tube of inner diameters ranging from 0.9 to 1.7 cm and lengths ranging from 10 to 18 cm. Some typical results obtained by Kandaurov et al.[15] from the study is shown in Fig.6-11. Using the first peak values of the diode voltage and beam current given in the figure, one can estimate the γ value which is about $\gamma = 2.5$. Further from Eq.(6-28) and using $I = 8.5$ kA, one can obtain the ratio R/r_2 to be about 1.37 which indicates that the beam radius is about 30% smaller than that of the tube.

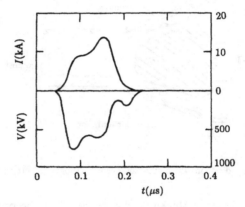

Fig.6-11 Typical results obtained from a foilless e-beam diode of similar configuration shown in Fig.6-9 [15].

6-3 Ion Beam Diode

(a) General Remarks

While high power electron beam has a wide range of applications, it also has limitations in certain aspects. For example, when the e-beam is employed in the particle beam pellet fusion research, the efficiency of the energy deposition is rather low. This is because the e-beam has a relatively long stopping range and can easily pre-heat the target. It also generates intense bremsstrahlung radiation during its interaction with the pellet. Both of these effects reduce the efficiency. Ion beams in these respects are more attractive and can yield higher efficiency in energy deposition. This is one of the major reasons that motivates the developments of high power ion beam source. Development of ion diodes is the heart of this work.

As noted in Sections 6-2(b), the bipolar flow of electrons and ions occurs in high current e-beam diodes. The principal effort in the development of high current ion diodes has been centered on the development of techniques for the suppression of the unwanted electron flow in the diodes. Various methods have been investigated for the suppression of electron flow and enhancement of ion current. Three methods were found to be effective and have been widely employed in practical cases. One of these methods is to make the electrons reflex back and forth between two cathodes thus reducing the electron flow. The device developed on this basis is referred to as reflex triode. The second method is, by means of magnetic field, to prevent the electrons from reaching the anode. This type of diode is called magnetically insulated ion diode. The third method is, by means of self-magnetic pinch, to enhance the impedance of the electron current via the increase of the electron path in crossing the diode gap. This type of diodes is termed as pinched ion diode. In the following we shall respectively examine each type of ion diodes in more detail.

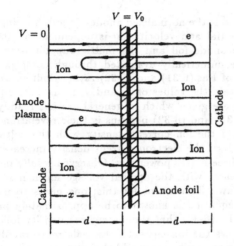

Fig.6-12 Idealized configuration and particle trajectories of the reflex triode.

(b) Reflex Triode.

The idealized configuration of the reflex triode is shown in Fig.6-12. It consists of an anode at high positive voltages placed between two cathode planes at zero potential. The anode consists of a metallic foil of a fractional electron range thickly covered on both sides by a surface plasma. The essential feature of the reflex triode is that electrons emitted from the cathode traverse the anode foil and subsequently see a reverse field which reflects the electrons back to the anode. As the electrons traverse the anode foil, they lose energy. The electrons continue to reflex through the anode until their energy becomes too low to penetrate the foil. Accumulation of those nearly stationary electrons near the anode, greatly enhance the electron density there which in turn increases the ion current density that can be extracted from the anode plasma. If the cathode is transparent, the ions will emerge from the back of the cathode and form an ion flow. In real cases, use of the second cathode is not necessary because the space charge beyond the anode will form a virtual cathode which can provide the reflection of electrons.

When steady state condition is assumed, the electron and ion current densities in a reflex triode may be expressed as[16]

$$J_e = \left(\frac{2e\epsilon_0^2}{m_0} \right)^{1/2} \frac{V_0^{3/2}}{d^2} \overline{L}^2 \tag{6-31}$$

and

$$J_i = (1 + \nu) \left(\frac{Zm_0}{M} \right)^{1/2} J_e \tag{6-32}$$

In these expressions, $\overline{L}$ is the normalized anode-cathode spacing and is an implicit function of the potential and velocity. ν is dependent on the applied voltage, the thickness of the anode foil and the multigroup diffusion coefficient. Both $\overline{L}$ and ν need to be numerically evaluated through complex procedures which make the usefulness of Eqs.(6-31) and (6-32) rather limited. In other words, one cannot expect to obtain the values of J_e and J_i from these equations unless the parameters $\overline{L}$ and ν are given which in general are not readily obtainable. One usefulness of Eqs.(6-31) and (6-32) perhaps is their revelation of the scaling law for the dependency of the ion current density with the voltage. According to the Child-Langmuir law, the electron current density increases with the applied voltage following a three-halves power law. Observed results indicate that the ion current density increases with the applied voltage much more rapidly than the three-halves power law. To account for this, one needs to use Eq.(6-32). The parameter ν in the equation is known to be approximately proportional to V_0^2. It can be seen from Eq.(6-32) that the ion current density follows a seven-halves power law which shows the ion current does scale more rapidly than the three-halves power law. If a second anode is added next to the first one, further increase in ion current density can be achieved. However, there are some criteria that have to be met before one can satisfactorily produce an ion beam from a reflex triode. These include the requirements of an magnetic field of 2–10 KG to prohibit the electrons expelled from the diode region and large gap spacing to prevent gap closure. In addition, large area of cathode surface is required to provide sufficient current.

Although the reflex triode can generate high ion current densities, in practice it has a number of shortcomings. Since the anode foil is subjected to intense electron bombardment it is usually destroyed in one shot. In general it also requires a strong axial magnetic field to maintain one-dimensional flow, however, angular momentum considerations make it impossible to extract parallel ion beams into free space. Finally the ion production efficiency η of the device is given by

$$\eta = \frac{J_i}{J_i + J_e} \approx \frac{1}{1 + \frac{1}{\alpha} \left(\frac{M}{m_0} \right)^{1/2}} \qquad (6\text{-}33)$$

At optimal condition, $\alpha \approx 11$, so the maximum efficiency is only about 20%, i.e. most of the diode current is contributed from electrons[11]. Variation of the ratio of ion current density to electron current density with α, where α is the average number of anode transits by electrons, is significant only when α is close to 11. Because of the various problems mentioned above, reflex triode has not been widely used in practical applications. It appears more appropriate to treat it as a development with new concepts rather than a practical device.

(c) Magnetically Insulated Ion Diode

As the reflex triode in real applications has a number of shortcomings a more practical method was developed to increase the production efficiency of intense ion beams from high voltage diode. In this method, a transverse magnetic field is applied across the diode gap. An idealized configuration of the magnetically insulated ion diode is shown in Fig.6-14. The underlying principle of the diode is that the magnetic field is sufficiently strong that the electrons emitted from the cathode will execute magnetron-like orbits and not be able to reach the anode plane. The minimum strength of the magnetic field required to produce such magnetic insulation is given by[17]

$$B_c = \frac{1}{d_0} \left[\frac{2m_0 V_0}{e} + \frac{V_0^2}{c^2} \right]^{1/2} \qquad (6\text{-}34)$$

B_c is usually referred to as the critical field and below which electrons will be able to reach the anode.

While the magnetic field confines the electrons to the cathode region, the more massive ions flow relatively unimpeded. If the cathode is transparent, intense ion beams can be extracted from the diode. In crossing the diode gap, the positive ions are also deflected by the magnetic field, but due to their massive mass, their deflection is quite small usually in the order of one to two degrees. Such angular deflection can be estimated by using the formula

$$\bar{\theta} = \sqrt{\frac{m_0}{M}} \frac{B}{B_c} \qquad (6\text{-}35)$$

where M is the mass of ions, B and B_c are respectively the applied insulating and critical magnetic fields. If the thickness of the electron sheath is much less than the gap spacing, i.e. $\delta \ll d_0$ and the ion deflection angle $\bar{\theta}$ is negligible, the ion current density in the magnetically insulated diode is then given by[18]

$$J_i = \frac{4\epsilon_0}{9} \left(\frac{2e}{M}\right)^{1/2} \frac{V_0^{3/2}}{d_0^2} \qquad (6\text{-}36)$$

The electron current density can be approximated by

$$J_e \approx \frac{\epsilon_0}{3} \left(\frac{2e}{\xi m_0}\right)^{1/2} \frac{V_0^{3/2}}{d_0^2} \qquad (6\text{-}37)$$

where $\xi \approx 3\delta/(2d_0)$. When $\delta \approx d_0$, i.e. when the electron sheath extends closely to the anode plane, the ion current density can be enhanced by about three times of the Child-Langmuir value given in Eq.(6-36). The above expressions were derived on the basis of a nonrelativistic diode with a strong transverse magnetic field applied. When these conditions are not satisfied, analytic formulas cannot be obtained and determination of the current densities will have to be done via a numerical method.

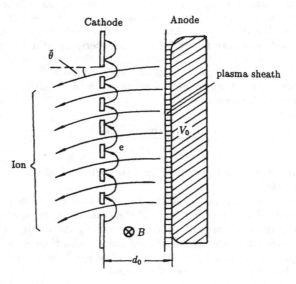

Fig.6-14 An idealized one-dimensional schematic diagram of the magnetically insulated ion diode with holes on the cathode.

There are two basic requirements for the satisfactory operation of a magnetically insulated diode, a plenty and prompt supply of ions at the anode surface and a properly designed magnetic field configuration. The former requirement can be relatively readily solved by using surface flashover technique[19], i.e. attaching a flashboard, such as aluminized plastic, to the anode surface. The breakdown along the dielectric surface produces the anode plasma which supplies the ion source. The magnetic field configuration required to produce the magnetic insulation is more complex, since it requires that field lines do not cross the diode gap. Various configurations of magnetically insulated diode have been designed in the past. One design using two parallel electrodes with the cathode having many holes in it. The ion beam is extracted through these holes as indicated in Fig.6-14. In this design the required field strength may be estimated from Eq.(6-34). When polarities of the electrodes shown in Fig.6-14 are reversed, negative ion beams can also be extracted from the device[20].

A second type of diode was developed by SNL[21] and a schematic of the diode is reproduced in Fig.6-15. The diode is cylindrically symmetric and produces a disk beam of positive ions. Fig.6-15(a) shows the cross sectional view of the left half of the diode structure. Two bundles of B-field coils, which provide the insulating magnetic field, are located above and below the disc cathodes. The anode-cathode region is separated from the gas-filled drift region by a Mylar foil. The applied magnetic field lines are shown in Fig.6-15(b). Electrons emitted from the tips of the cathode disks spiral axially along magnetic field lines to form a virtual cathode in front of the anode surface. Surface flashover along a Nylon mesh mounted on the anode produces an anode plasma which serves as the ion source. To achieve efficient operation of the diode, it requires that the diode voltage must be below the critical voltage V_c which is given by

$$V_c = m_0 c^2 \left\{ \left[\frac{e\Psi(r_a) - e\Psi(r_c)}{r_a m_0 c} \right]^2 + 1 \right\}^{1/2} - m_0 c^2 \qquad (6\text{-}38)$$

In the above expression, r_a and r_c are respectively the position vectors of the anode and cathode, $\Psi(r)$ is the magnetic stream function defined as

$$\Psi(r) = r A_\theta(r)$$

where A_θ is the azimuthal component of the vector-potential and is related to the current density J by

$$\epsilon_0 c^2 \nabla^2 A = -J$$

When the diode voltage is above V_c, magnetic insulation of the diode gap will not be maintained. The ion current density that can be extracted from the diode in general is greater than the Child-Langmuir value given by Eq.(6-36) and how much greater is determined by many factors. There is no analytic formula available for the calculations of current densities which have to be done via a numerical method.

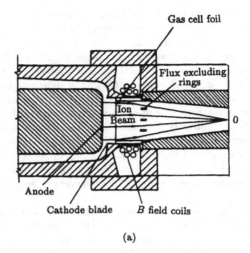

(a)

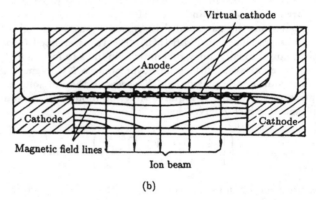

(b)

Fig.6-15 Schematic of a magnetically insulated ion diode [21]. (a) Basic structure of the diode, (b) Illustration showing the magnetic lines, electrons and ions trajectories.

By using the magnetically insulated ion diode, LiF ion source experiments have been conducted recently on the PBFA-II accelerator in the USA[22]. The anode radius, i.e. the distance from point 0 to the anode surface shown in Fig.6-15(a), is about 15 cm. The anode-cathode gap is between 1.35 to 1.65 cm. The gas cell foil is a 2 μm Mylar which is located at 12.5 cm from the center line. The drift region is filled with Ar gas of 1 torr pressure. When a 10 MV, 20 ns rise time voltage pulse applies at the diode, a virtual cathode is formed in front of the anode and the process can be readily described by existing theories[23]. Measurements carried out by these authors indicate that, after the beginning of ion emission, the electric field in front of the anode is about 8 MV/cm. The Li$^+$ ion current density so obtained is about 1–2 kA/cm^2 and the corresponding voltage is 6–9 MV. The current wave forms may be approximately represented as follows (Fig.6-16).

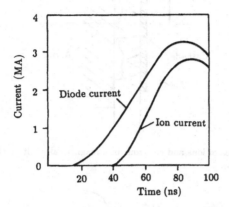

Fig.6-16 Schematic of the current wave forms obtained by SNL researchers from a magnetically insulated ion diode of similar configuration shown in Fig.6-15 [22].

(d) Pinched Ion Diode

As described in Section 6-2(b) and illustrated in Fig.6-8(d), in a large aspect-ratio diode, a quasi-steady state bipolar flow can be established as a result of the tight pinch of the electron beam. The total diode current consists of a laminar ion flow and a pinched electron current. The ratio of ion to electron currents can be approximately expressed by an analytic expression[25]

$$\frac{I_i}{I_e} \geq \frac{R}{d_0} \left(\frac{eV_0}{2Mc^2} \right)^{1/2} \tag{6-39}$$

If the aspect ratio R/d_0 is sufficiently large, equal ion and electron currents is possible. Such enhancement of ion current is the result of particle paths difference. In the diode gap, electrons follow a wiggly long path whereas ions follow an almost straight line path as shown in Fig.6-17. Based on this principle, several types of pinched ion diode have been developed in recent years. The most notable ones are the pinch-reflex diode[26,24], the inverse pinch diode[27] and the point pinch diode[28].

Fig.6-18 shows the configuration of a standard pinch-reflex ion diode. The cathode tip is in annular form and having a thin foil associated with it. The

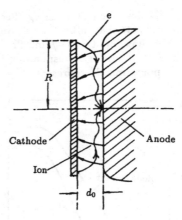

Fig.6-17 Trajectories of ions and electrons in a magnetically self pinched ion diode of large aspect ratio.

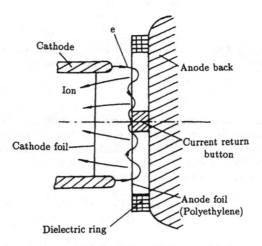

Fig.6-18 Schematic of a standard pinch-reflex ion diode.

anode-cathode gap in this case is 5 mm. The anode consists a 12.7 μm-thick polyethylene foil stretched taut across a 5 mm thick dielectric ring at a diameter larger than the cathode diameter. A thin (178 μm) aluminum annulus with an inner diameter slightly smaller than that of the cathode is placed on top of the plastic foil across from the cathode tip. Such a measure is to ensure the electron beam pinch well-behaved and symmetrically collapsing. The vacuum gap between the anode foil and the anode back-plate is 5 mm and the current is returned via a central, 6.4 mm diameter button as shown in the figure. Electrons emitted from the cathode tip, reflex through the plastic foil as a result of their interactions with the azimuthal magnetic field established by the current flowing through the central button. The ions extracted from the anode plasma are accelerated across the diode vacuum gap, pass through a thin (2 μm) polycarbonate foil and are transported toward the target in gas cell (1 torr air). It was noted that a significant quantity of debris was created during a shot due to the interaction of the electron beam with the anode structure, particularly the central button and the solid anode backing plate. In order to alleviate this problem, a modified version of the diode was recently investigated by Stephanakis et al[26].

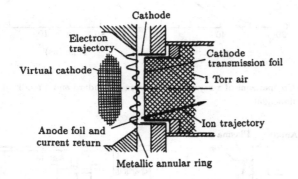

Fig.6-19 Schematic of a low-debris pinch-reflex ion diode [26].

The configuration of the modified diode (low debris version) is reproduced in Fig.6-19. In this version, the central anode button and the anode backing plate are completely eliminated. There is a large evacuated region behind the anode foil so that the electrons are allowed to perform large orbits until they are reflected back by a self-generated virtual cathode. Upon such modification, X-ray yield and ion efficiency of the diode were found to be significantly improved. Fig.6-20 is one of the results obtained by the same authors[26] which indicates that the modified diode generates about 2.5 times the X-ray per unit time than the same diode backed with a thick carbon block. The experiments described above were carried out on the Gamble II generator in the USA. The typical operation parameters were 1.3 MV, 0.85 MA and 1.1 TW. The ion beam consists of mainly protons with a small percentage of carbon ions.

Proper focusing of intense ion beams extracted from the diode is one important yet difficult task. Inverse pinch ion diode offers a relatively simple method to focus the ion beams[29]. The basic structure of the inverse pinch diode and the essential experimental set-up employed by Hashimoto et al.[27] are shown in Fig.6-21. The diode consists of an annular anode and an annular cathode projected

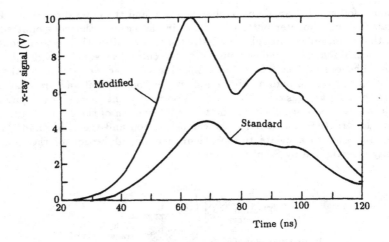

Fig.6-20 Comparisons of x-ray output between a standard and a modified pinch reflex ion diode [26].

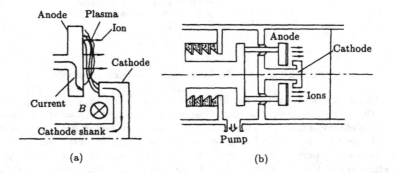

Fig.6-21 (a) Schematic drawing of the "Inverse Pinch Ion Diode", (b) the experimental setup [27].

from a circular disk which is connected to the ground plane via a cathode shank as shown. Electrons emitted from the cathode tip are accelerated toward the anode. The current flowing in the cathode shank generates a strong azimuthal magnetic field which provides the essential condition for magnetic insulation and formation of a virtual cathode in front of the anode. Ions emitted from the anode plasma are accelerated essentially along the electrostatic field lines and focus on the diode axis. The outer diameters of the cathode disk and the anode ring are

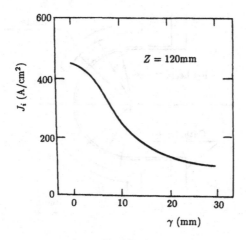

Fig.6-22 The radial profile of the ion beam at 12 cm on the diode axis, where J_i is the ion current density, z is the axial distance from the anode and r is the radial distance from the axis [27].

respectively 1.5 and 5 cm. A 2 mm thick acrylic plate was attached to the anode surface as a plasma source. Results obtained by the same investigators indicate that the ions are focused on the diode axis with a maximum ion current density of 0.5 kA/cm^2 at 12 cm distance from the anode surface. The focusing radius of the ion beam (FWHM) is between 1.0 to 1.5 cm as can be seen in Fig.6-22. The peak value of the ion current is 5 kA which gives an ion generation efficiency (I_i/I_d) of 5%. In comparison with other diodes, such an ion efficiency is rather low which represents one major disadvantage of the inverse pinch diode.

(e) Plasma Focus Diode

In many applications, such as in the Inertial Confinement Fusion (ICF) research, the ion beam is required to be focused on a small area less than a square mm. The focusing radius achieved by the inverse pinch diode described in the last section is far too large to meet this requirement. In the magnetically insulated diodes described in Section 6-3(c), due to the complex configuration of applied magnetic field and instabilities developed in the gap, further reductions of the beam divergence is difficult. By using a spherical plasma focus diode developed recently[29], some investigators have obtained beam focusing radius of about 200 μm at the focusing area.

The basic structure of the spherical plasma focus ion diode employed by Jiang et al.[29] is in many respects similar to the self magnetically insulated diode[24] mentioned at the end of Section 6-3(c). It consists of a pair of concentric spherical

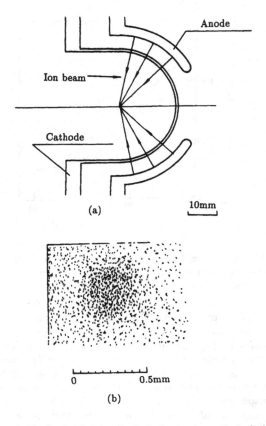

(a)   10mm

(b)

Fig.6-23 Self magnetically insulated, spherical plasma focus diode [29]. (a) The basic structure; (b) Photograph of ion beam focused on an aluminum foil.

electrodes as shown in Fig.6-23 (a). The radii of the anode and cathode are 2.5 cm and 2 cm, respectively, in this case. Grooves of 0.8 mm wide, 1 mm deep and 1.5 mm in pitch are trenched in the azimuthal direction on the anode surface which are filled with epoxy. The cathode is perforated with 1 mm holes and having a transparency of approximately 40%. Electrons emitted from the cathode are guided by the azimuthal magnetic field generated by the diode current to drift along the curved gap. Analytic calculation have shown that most electrons drift out of the diode gap at the downstream end before reaching the anode[30]. Thus the diode is self magnetically insulated. Fig.6-23(b) is a photograph obtained by Jiang et al.[29] from a CR-39 detector. In the experiment, the ion beam was focused on a 7 μm thick aluminum foil located at the spherical center of the electrodes.

The spatial distribution of the ion density read from the figure indicates that the FWHM of the density profile is about 0.4 mm in all directions which gives a divergence angle of about one degree in the azimuthal direction. When the aluminum foil was moved 2 mm away from the center, a clear picture of the beam profile is no longer obtainable indicating that the ion beam was sharply focused at the center position. Based on the observed results, the investigators have estimated the ion current density at the focused position to be about 680 kA/cm^2 with a power density of 0.54 TW/cm^2 and ion energy of 0.8 MeV.

References of Chapter 6

[1] R. H. Fowler and L. W. Nordheim, *Proc. Roy. Soc. Lodon*, **A119**, (1928), 173.

[2] F. Rohrbach, *CERN Report*, 71-5/TC-L, (1971).

[3] W. W. Dolan, W. P. Dyke and S. K. Trolan, *Phys. Rev.*, **91**, (1953), 1054.

[4] G. A. Mesyats and D. I. Proskurovsky, *Pulsed Electrical Discharge in Vacuum*, (Springer-Verlag, Berlin, 1989), p.85.

[5] Qi Zhang et al., *Rev. Sci. Instrum.*, **62**, (1991), 1658.

[6] J. E. Boers and Kelleher, *J. Appl. Phys.*, **40**, (1969), 2409.

[7] L. A. Rosocha and K. B. Riepe, *Fusion Technology*, **11**, (1987), 576.

[8] J. M. Creedon, *J. Appl. Phys.*, **46**, (1975), 2946.

[9] S. A. Goldstein et al., *Phys. Rev. Lett*, **33**, (1974), 1471.

[10] S. A. Goldstein et al., *1st Topical Conf. Etect. Beam Res. and Technol.* Sandia National Lab., SAND 76-5122, (1975), 218.

[11] R. B. Miller, *An Introduction to the Physics of Intense Charged Particle Beams*, (Plenum Press, New York, 1982).

[12] R. B. Miller et al., *J. Appl. Phys.*, **51**, (1980), 3506.

[13] Wang Ganchang, *High Power Laser and Particle Beams* (in Chinese), 1, (1989), 1.

[14] N. G. Tsien et al., Institute of Atomic Energy of China, Tech. Report (in Chinese), (1991 May).

[15] I. V. Kandaurov et al., *9th Int. Conf. on High Power Particle Beams*, (1992), 1027.

[16] T. M. Antonsen and E. Ott. *Appl. Phys. Lett.*, **28**, (1976), 424.

[17] T. M. Antonsen and E. Ott, *Phys. Fluid*, **19**, (1976), 52.

[18] J. Gold, T. J. Orzechowski and G. Bekefi, *J. Appl. Phys.*, **45**, (1974), 3211.

[19] S. Humphries et al., *J. Appl. Phys.*, **47**, (1976), 2382.

[20] K. Horioka et al., *9th Int. Conf. on High Power Particle Beams*, **II**, (1992), 806.
J. A. Nation, *Particle Accelerators*, **10**, (1979), 1.

[21] S. A. Slutz, D. B. Seidel and R. S. Coats, *J. Appl. Phys.*, **59**, (1986), 11.

[22] R. W. Stinnett et al., *9th Int. Conf. on High Power Particle Beams*, **II**, (1992), 788.

[23] M. P. Desjarlais, *Phys. Fluid B*, **1**(8), (1989), 1709.

[24] G. Cooperstein, *Laser Interaction and Related Plasma Phenomena*, **5**, (Plenum Press, New York, 1981), p.105.

[25] S. A. Goldstein and R. Lee, *Phys. Rev. Lett.*, **35**, (1975), 1079.

[26] S. J. Stephanakis et al., *9th Int. Conf. on High Power Particle Beams*, (1992), 871.

[27] Y. Hashimoto et al., *8th Int. Conf. on High Power Particle Beams*, (1990), 475.

[28] M. Soto and T. Tazima, *8th IEEE Pulsed Power Conf.*, (1991), 578.

[29] W. Jiang et al., *9th Int. Conf. on High Power Particle Beams*, (1992), 859.

[30] W. Jiang et al., *8th Int. Conf. on High Power Particle Beams*, (1990), 733.

CHAPTER 7
APPLICATIONS OF HIGH
POWER PULSE TECHNOLOGY

7-1 Particle Beam Generator

In the preceding chapters, we have discussed the basic principles and functions of the various components and devices that are essential for the design and operation of a high power pulse system. Now we are in a position to discuss how to put these components and devices together to form a complete, functional system from which to generate the desired results — in this case the high power particle beams. When a high power pulse system is designed to generate particle beams, other than the name of particle beam generator, it is frequently referred to as a pulsed power generator to imply wider applications. In this chapter, we shall adopt both conventions to naming the system.

Table 7-1 Pulsed Power Generators of one TW Class

Generator	Type of PFL	Diode Voltage (MV)	Diode Current (MA)	Output Power (TW)	Pulse Width (ns)	Location and year
HERMES-II	Oil Blumlein	10	0.1	1	80	USA 1968
GAMBLE-II	Water Coaxial	1	1	1	50	USA 1970
OWL-II	Water Coaxial	1.3	0.8	1	120	USA 1975
ANGARA-II	Water Coaxial	1	1	1	60	Russia 1976
FLASH-I	Oil Blumlein	8	0.1	0.8	80	China 1979
SIDONIX-II	Water Coaxial	1	1	1	105	France 1980
FLASH-II	Water Coaxial	1.2	0.6	0.7	89	China 1988

In the past 30 years or so, there have been many types of pulsed power generators developed around the world and the power level generated has been multiplied many times. For example, the most advanced generator, PBFA-II of the USA, is now capable of producing pulsed power at 100 TW level and generating 5 MeV, 18 TW proton beams[1]. Such advanced generators involve rather complex design and structure. For illustration purposes, we shall in this chapter use some generators having relatively simple structures and less advanced designs to illustrate the general characteristics and essential features of the pulsed power generators. Table 7-1 lists some of the pulsed power generators in the one-TW category. These generators have relatively simple structures and are therefore more suitable for discussion.

From Table 7-1, one can see that there are two types of pulse forming lines (PFL), oil Blumlein lines and water coaxial lines, that have been commonly employed for this type of pulsed power generators. One may also notice from the table that the trend of generator developments is clearly in favor of the water line over oil line, as nearly all the later designs were using water coaxial lines. There are a number of reasons which make the water line more attractive[2]. Perhaps its suitability for low impedance diodes is one of the major reasons, as under a given voltage, the output power is inversely proportional to the impedance of the diode. This is, however, by no means implying that the technology of oil transmission lines can be neglected. On the contrary, the oil transmission line is still technically important and useful, particularly when a high impedance diode is required. For this reason, we shall use a generator built on the basis of oil line technology as our first example to illustrate how a pulsed power generator works and what it can do.

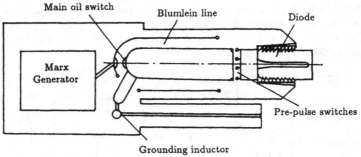

Fig.7-1 Schematic of the pulsed power generator FLASH-I [3].

Fig.7-1 is a schematic of the generator Flash-I built by China Academy of Engineering Physics (CAEP)[3]. This generator was built on the basis of the oil Blumlein line technology and designed chiefly for the production of intense γ-ray via the process of e-beam interaction with target material[3]. On the leftmost of the figure, the block represents a Marx generator which serves as the energy source and voltage multiplier. It consists of 102 high voltage capacitors of 100 kV and 0.63 μF each. When these capacitors are fully charged, the output voltage of the Marx generator was designed to have a peak value of 10 MV. In actual operations, however, the peak values of the voltage are generally between 7.5-8.4 MV. Next to the Marx generator is a coaxial Blumlein line consisting of three coaxial cylinders with one spark gap switch connected between the inner and middle cylinders. The load (diode) is connected between the inner and outer cylinders. The Blumlein serves as a pulse forming line, as the width of the voltage pulse generated from the Marx generator usually is in the microsecond range and it is too slow to be useful for the application under consideration. The Blumlein line is capable of compressing the voltage pulse from the Marx generator into a well shaped pulse of fast rise time and pre-determined width. The pulse width can be determined from the length of the Blumlein line and the dielectric constant of liquid employed. The inductor connected to the inner cylinder of the Blumlein is to provide a grounding path for the otherwise floating inner cylinder to be charged during the charging phase. This is because the diode is essentially an open circuit during the charging phase. Without the inductor, the inner cylinder would be floating and cannot be properly charged. However, with the inductor connected, there will be current leaking away through it hence energy loss. Therefore the selection of proper values for the inductor is important.

At the right end of inner cylinder, there is a set of pre-pulse switches whose function is to reduce the effects of the pre-pulse voltage on the diode. Pre-pulse voltage is the result of capacitive coupling between the PFL and diode. As switches and diodes are essentially capacitors, during the charging phase of the PFL a pre-pulse voltage appears at the diode before the main voltage arrives. Such an effect can cause erratic diode performances such as: (1) early collapse of diode impedance, (2) unnecessary loss of energy, (3) improper pinching and focusing of the e-beam. Several techniques have been developed to reduce the pre-pulse effects[4,5]. A common method is to separate the transmission line into sections and place a switch between them as that shown in Fig.7-1. If the pre-pulse switch is properly adjusted such that it closes at a proper time before the main voltage pulse arrives, substantial reduction of the pre-pulse effects can be achieved. However, for a generator with a vacuum diode as the load, it is more effective to use a plasma erosion switch to short the diode during the charging phase. This way practically isolates the diode from the charging circuit and therefore can greatly reduce the effects.

The vacuum diode in this case is a high impedance diode. It was designed to generate well focused electron beams which require the cathode tip area to be relatively small. From Eq.(6-11) and Eq.(6-17) we can see that such a diode has a relatively large impedance and a small current in comparison with that of other generators listed in Table 7-1. The entire structure of the Flash-I generator, except the vacuum diode, is enclosed in a tanker which is filled with high voltage oil. The oil provides the necessary insulation for the structure as well as the dielectric medium for the Blumlein line. The basic principle of operation of the Flash-I generator and the observed outputs may be qualitatively described by using the block diagram shown in Fig.7-2. Upon firing the Marx generator, a voltage pulse of approximately 8.4 MV peak, 1.4 μs rise time is generated. At about the peak value of the voltage pulse, the main switch S_1 closes and a fast voltage pulse of about 85 ns duration is thus formed and the Blumlein line is charged. This fast pulse propagates toward the diode, when it arrives there is an e-beam pulse of similar nature generated from the diode. Interaction between the e-beam and anode material, such as Tantalum can produce intense γ-radiation. If the high-Z target is made less than one electron range thick, some electrons can pass through the high-Z target and be absorbed in a low-Z target as shown in Fig.7-2. The parameters of operation and the characteristics of the various outputs from the generator are summarized in Table 7-2, and Fig.7-3.

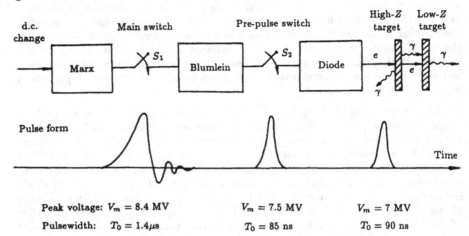

Fig.7-2 Block diagram showing the operating principle and observed outputs of the Flash-I generator.

Table 7-2 Operation Parameters and Output Characteristics
of Flash-I generator (3).

Marx Charged Voltage (kV)	45	60	70	80	85
Main Switch Gap (cm)	10	15	18	20	21
Diode Gap (cm)	14	14	18	20	20.3
Diode Voltage (MV)	3.8	5.0	5.9	7.5	8.0
Diode Current (kA)	43.9	64	77.6	92.3	93.5
γ-Ray Dose (R)*	106	334	569	819	961
γ-Dose Rate ($\times 10^{10}$ R/S)*	0.1	0.4	0.9	1.6	2.1

* Measured at 1m from the target 1 $R = 2.58 \times 10^{-4}$ C/kg.

Fig.7-4 is a schematic of the generator Flash-II built by Northwest Institute of Nuclear Technology (NINT)[6]. The basic structure of this generator is similar to that of the Flash-I generator, namely it consists of basically a Marx generator, a coaxial line and an e-beam diode[6]. The main differences are that this generator is built on the basis of water coaxial line and having low impedance diode whereas in the Flash-I generator, oil Blumlein line, and high impedance diode were used. As the output power is inversely proportional to the diode impedance, Flash-II has the advantage in achieving same power with significantly less diode voltage. For this reason, the Marx generator in this case was constructed with 64 capacitors of similar specifications, yet 102 capacitors were required in Flash-I. Moreover, a

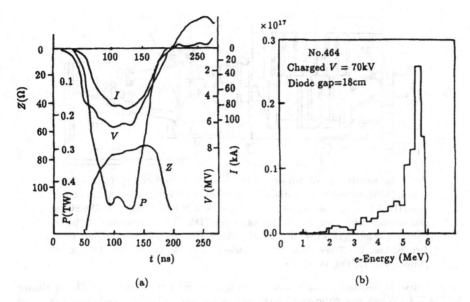

Fig.7-3 Outputs characteristics of Flash-I generator obtained by researchers at CAEP [3]. (a) Current I, Voltage V, Power P and impedance Z; (b) Electron energy spectrum.

lower voltage implies less requirements for insulation consideration thus enabling the generator's structure to be more compact and smaller than that of the Flash-I. For example, the physical lengths are 17 m for Flash-II versus 26 m for Flash-I. The main disadvantage of the water coaxial line is that the maximum output voltage is only half of the charging voltage, i.e. only half of the voltage generated from the Marx generator is available for power production whereas in the case of oil Blumlein line nearly the full amount of voltage can be utilized if the impedances are properly matched.

The diode in this case has a rather large aspect ratio and its nominal impedance is designed to be $Z_D = 2\Omega$. In order to match the diode impedance the output line (OL) is designed to have the same value of impedance, namely $Z_0 = 2\Omega$. The other two parts of the coaxial line are the pulse forming line (PFL) and the transmission line (TL). The function of the former is to compress the voltage pulse into a narrow one and the latter is to transmit and further sharpen it. The impedances of the PFL and TL are 5 Ω and 3.2 Ω, respectively. Between the PFL and TL is a water switch serving as the main output switch. The basic nature of this switch is similar to the multichannel water switch described previously in Section 5-3 and illustrated in Fig.5-26. It consists of essentially three parts, a trigger disk, a gas spark gap and an energy absorber or an inductor connected to ground. The

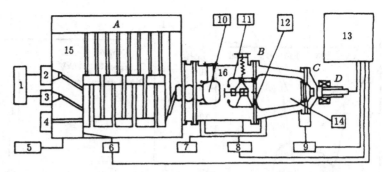

Fig.7-4 Schematic of FLASH-II accelerator system built by NINT [6]. (a) **Marx gen-erator**; (b) **Water-dielectric coaxial line**; (c) **Diode**; (d) **Pulse magnetic field and drift room (B=1.8 T)**. 1. Control system, 2. Trigger, 3. H.V.Power supply, 4. SF6 Filling and venting installation, 5. Oil filtering system (180 T), 6. I_M Monitor, 7. Water purifying system (15 T), 8. U_F, U_r, U_{01} Monitor, 9. I_D, U_{02} Monitor, 10. Pulse forming line, 11. Trans. line, 12. Pre-pulse switch, 13. 120 dB Shield room, 14. Output line, 15. Oil, 16. Water.

trigger disk is located outside the TL and the rest are within the TL as shown in Fig.7-4. When the concerned parameters are properly adjusted, multi-channel breakdown between the PFL and TL which can considerably reduce the system's inductance will occur. There are 8 gas switches arranged in parallel configuration and located between the TL and OL. The function of these switches is to reduce the pre-pulse effects as discussed previously. The other parts of the generator are explained in Fig.7-4 and the functions of those parts can be identified there. Table 7-3 gives the operation parameters and outputs characteristics of the Flash-II generator. Some of the observed results obtained by Qiu Aici et al.[6] from the Flash-II generator are shown in Fig.7-5 to Fig.7-7. A comparison of these results with that shown in Fig.7-3, can show that there are many similarities between the two results indicating that the Flash-II generator has comparable capability inspite of its relatively small size and low charging voltage.

Table 7-3 Operation parameters and outputs characteristics of Flash-II generator (6).

Charge voltage V_0 (kV)	55	70	80	85
Main switch distance d, (mm)	120	151	161	171
Cathode radius R (mm)	90	110	110	110
A-K gap d (mm)	11–12	8–8.5	8.1	8.1
Diode voltage V_D (kV)	936	1072	1302	1638
Diode current I_D (kA)	196	466	589	511
Total beam energy ϵ (kJ)	11	30.5	52.2	66
Diode impedance Z_D (Ω)	5.1	2.1	2.1	3.1

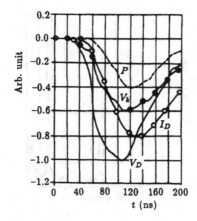

Fig.7-5 Diode voltage V_D (peak 1302 kV), A-K voltage V_k (peak 1113 kV), current I_D (peak 589 kA) and power P (peak 6.2×10^{11} AV) obtained by Qiu Aici et al. from Flash-II generator [6].

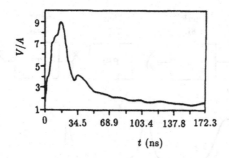

Fig.7-6 Diode impedance obtained by Qiu Aici et al. from Flash-II generator [6].

An electrical pulse can be progressively compressed and its rise time shortened by successively transferring the energy into stages of low inductance and capacitance. Most of the large pulsed power generators of advanced design utilize this method to reshape the voltage pulse generated by the Marx generator. Quite often the same method is also employed to achieve impedance match at the output end. Fig.7-8 illustrates a typical example of the method used in a generator design. In this example the successive stages include an intermediate store (IS), a pulse forming line (PFL), a transmission line (TL) and an output line (OL). With the assigned values of parameters labelled and proper selection for other conditions, calculations indicate that the compression of the pulse at each stage can be ex-

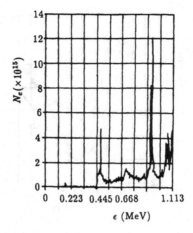

Fig.7-7 Electron energy spectrum obtained by Qiu Aici et al from Flash-II generator [6]. Mean electron energy 8.52×10^5 eV. Total beam energy 3.25×10^{23} eV.

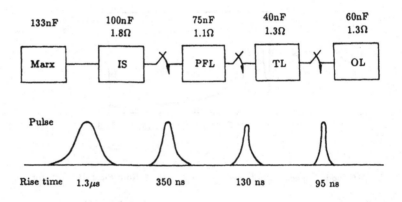

Fig.7-8 Schematic illustration of pulse compression and impedance matching by successive transfer of energy. IS=intermediate store, PFL=pulse forming line, TL=transmission line, OL=output line.

pected to be approximately as that shown in Fig.7-8. One may notice from the figure that the impedances of the TL and OL are also properly matched which is essential for achieving a well defined output pulse. The basic requirement for applying this method successfully is the use of proper values for the various parameters involved. Because the complexity of the system studied, use of the ordinary circuit theory is no longer adequate to find satisfactory answers for these values and transient transmission line theory is frequently needed. A common approach is to use some code or algorithm to simulate the behaviors of the system numerically. In the approach, a simulation model similar to that shown in Fig.7-9 is first constructed from available data, then the performance of the system under study is numerically simulated by means of the code. Various computer codes such as the JASON Code[7], MAGIC Code[8], SCREAMER Code[9] and GOST Code[10] etc. for such purposes have been developed by previous authors. Discussion of these codes is beyond the scope of this book. Interested readers are suggested to look up the appropriate references.

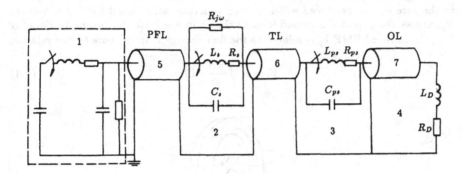

Fig.7-9 Simulation model for FLASH-II generator [6]. 1. Marx generator, 2. Output switch, 3. Prepulse switch, 4. Diode, 5. $Z_F = 5.0\Omega$, $T_F = 40$ ns, 6. $Z_T = 3.2\Omega$, $T_T = 40$ ns, 7. $Z_0 = 2.0\Omega$, $T_0 = 60$ ns.

7-2 Linear Induction Accelerator

Electron beams with both high energies and currents have a wide range of applications such as in the areas of free electron lasers, radiography and nuclear effects simulation etc. The particle beam generators discussed in the last section are capable of producing electron beams of quite high power and current, but the beam peak energy they can produce is limited below 15 MeV. Any attempt to increase the beam energy, would have to substantially increase the output voltage of the Marx generator which with the present technology is very difficult to achieve. An alternative is to employ the linear induction accelerating scheme in which the charged particles are accelerated by induced electrical fields in successive stages. As the number of accelerating stages can be many, the beam energy can be multiplied without the necessity of using very high voltages. Accelerators built on this basis and the accelerating stages are arranged in tandem form are called Linear Induction Accelerators (LIA). There are three types of LIA: the

Astron-type[11], Radlac-type[12] and Auto-accelerator devices[13]. The Astron-type accelerators use ferromagnetic material in the cores to generate the accelerating field whereas the Radlac and Auto-accelerators use air core cavities. In the Auto-accelerator the air core cavities are excited by the fields of electron beam itself, rather than by external fields. There are advantages and disadvantages associated with each type of accelerators. Selection of the proper type is largely determined by the requirements and applications in question.

The basic principle governing the induction accelerators may be understood via the illustration shown in Fig.7-10. Consider a toroidal core of ferromagnetic material which is surrounded by two single-loops with open gaps. When a driving voltage pulse of peak V_0 is applied to one of the loops, a change of magnetic flux B in the core will be produced which in turn generates an induced EMF, i.e. voltage V_g, across the gap of the second loop as shown in Fig.7-10. According to Faraday law, the induced EMF V_g is related to the flux change $\dot{B}$ in the core by the relation

$$V_g = -\int \frac{dB}{dt} \cdot dS \qquad (7\text{-}1)$$

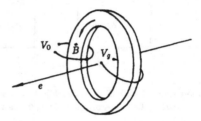

Fig.7-10 Schematic showing the basic principle of induction accelerator.

In the expression, the surface integral is taken over the cross section of the ferromagnetic core. The voltage V_g will be maintained in the gap until either the driving voltage pulse terminates or the magnetic material in the core reaches saturation. During the time V_g appears, a charged particle passing by the V_g gap will be accelerated as indicated in the figure. In a LIA, several accelerating gaps are arranged in tandem and synchronized to provide an accelerating electric field when the charged particles pass the gaps. Fig.7-11 shows the schematic of three modules (accelerating stages) of a LIA designed on the basis of Astron-type and by the China Academy of Engineering Physics (CAEP)[14]. From now on this accelerator shall be referred to as CAEP-LIA. In this design, Blumlein pulse forming lines are used to provide the driving voltage pulses for the ferromagnetic cores and the Blumlein lines are powered by a Marx generator. Upon closing the main switch on the Blumlein, the output voltage pulse from the Marx generator is re-shaped into a flat top pulse of 235 kV peak and 90 ns width as indicated in the figure. This well-shaped pulse drives the magnetic core to generate the accelerating voltage V_g across the gap. Synchronization of the electron beam with the accelerating fields is done by properly adjusting the closing time of each switch on the Blumlein lines. The accelerating voltage pulse so produced across each gap has the characteristics of 235 kV peak and 90 ns duration.

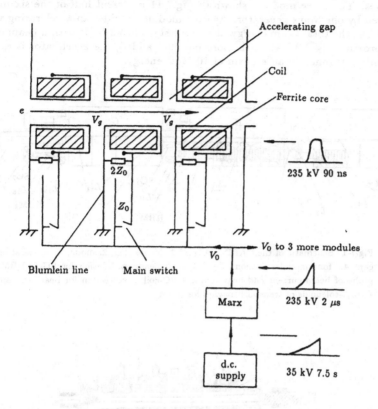

Fig.7-11 Schematic of the 3.3 MeV LIA showing three modules only. The waveforms from bottom to top are respectively the input and output of the Marx generator and the output of the Blumlein.

The entire structure of the CAEP-LIA consists of 12 accelerating modules and one drift tube for beam modulation. The 12 modules are divided into three sections A, B and C and each section contains 4 modules as shown in Fig.7-12. Section A serves as a beam injector whose structure mainly consists of a long cathode stalk extends the length of the four modules and with some velvet material stuck at its end as electron emitter. The anode is made from tungsten mesh as shown in Fig.7-13. At relatively low field strengths, electron emission from the cathode stalk is not serious, current can flow along the stalk surface and voltage addition occurs there[15]. As a result, the electrons emitted at the cathode tip can gain the full amount of energy supplied by the four accelerating gaps. The 12 modules shown in Fig.7-12 are powered by two Marx generators and each Marx shares six modules. The three modules shown in Fig.7-11 represent half of the six modules powered by one Marx generator. As each module provides an accelerating voltage of 235 kV, the total beam energy at present stage is 3.3 MeV with a beam current of approximately 2.2 kA. With more modules added, the accelerator is designed to eventually generate an e-beam of 10 MeV energy.

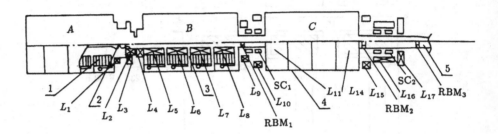

Fig.7-12 Schematic of the CAEP-LIA. (14) 1. ferrite ring; 2. diode; 3. accelerating gap; 4. four modules section; 5. section of beam modulation for SG-1; RBM-probe of beam current and beam station; SC-coil of correction for beam station; L_1-L_{17}-solenoids of transport magnet for beam

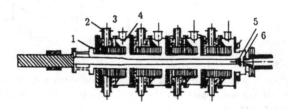

Fig.7-13 Schematic of section A in Fig.7-12. (14) 1. Ie monitor; 2. oil; 3. pump; 4. vacuum; 5. cathode; 6. anode.

From Figs.7-11 and 7-12, one can see that proper synchronization between the electron beam and the accelerating fields is critically important in operating such an accelerator. That is each switch in the entire system must function precisely

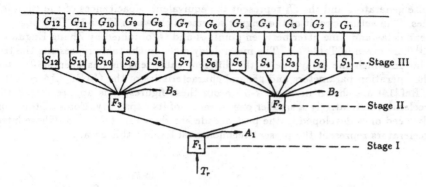

Fig.7-14 Block diagram of trigger scheme of the CAEP-LIA showing the sequence of 3-stage switching [14]. $T_r =$ initial trigger signal; $F_1 =$ first stage switching; F_2, $F_3 =$ second stage switching; $S_i =$ main switching, 3rd stage; $G_i =$ accelerating gap.

according to a pre-determined time sequence. Any one misfire may seriously jeopardize the beam characteristics. The trigger scheme employed in the CAEP-LIA is 3-stage switching as shown in Fig.7-14 by block diagram. When the first

Table 7-4 The mean delay and jitter between point A and G_i [14]

observing points	G_1	G_2	G_3	G_4	G_5	G_6	G_7	G_8	G_9	G_{10}
mean delay $\bar{A}_{A-G}$/ns	359.6	359.9	361.2	359.1	362.4	361.0	363.3	367.7	355.6	357.7
jitter σ_{A-G}/ns	2.8	2.3	1.5	1.9	2.4	1.7	2.4	3.1	3.4	3.3

G_{11}	G_{12}	B_2	B_3
364.0	357.6	152.7	131.5
3.5	2.6	1.4	1.3

stage switch F_1 is triggered, two trigger pulses are sent respectively to the second stage switches F_2 and F_3. Upon closing F_2 and F_3, 12 trigger pulses are further generated to trigger the 12 main switches on the Blumlein lines. By choosing proper length for each cable, synchronization between the electron beam and the accelerating field at each gap can be achieved provided all elements in the trigger scheme function properly. A conceptual equivalent circuit of the trigger scheme is shown in Fig.7-15 in which the C'_s represent the capacitances of the various pulse generators and the C_s represent the equivalent capacitances of the Blumlein lines. The actual circuit of the trigger scheme is shown in Fig.7-16. The observed mean delay and time jitter between points A and G_i obtained by Zhang Enguan et al.[14] are given in Table 7-4. The mean delay given in the table represents the true delay in which the delay due to the different cable lengths has been subtracted. The operation parameters and general characteristics of the CAEP-LIA reported in Ref.[14] are shown in Table 7-5. From these figures, it is apparent that this accelerator is relatively a smaller one in terms of its capability. Some of the more advanced ones developed in the past decade are shown in Table 7-6. These latter accelerators represent the present technological level in this area.

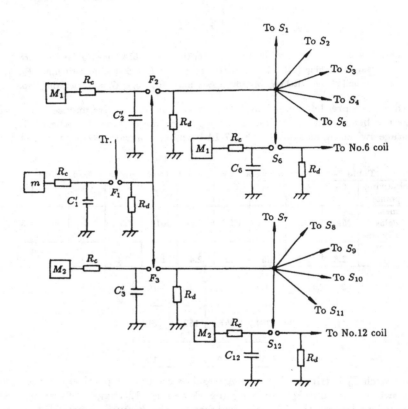

Fig.7-15 Conceptual equivalent circuit of the trigger scheme showing in Fig.7-14. m=mini Marx, M=main Marx, R_c=charge resistor, R_d=discharge resistor, F, S=switches.

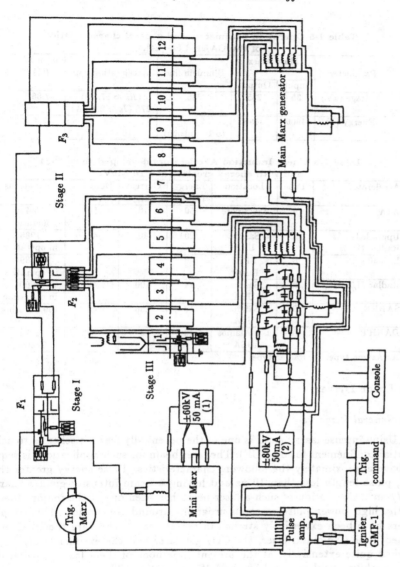

Fig.7-16 Actual circuit diagram of the trigger scheme employed in the CAEP-LIA [14].

Table 7-5 Operation parameters and general characteristics of the CAEP-LIA [14].

Parameters	Main Marx		Blumlein line	Accelerating gap	η_r or η_e (%)
	Input	Output			
Voltage (kV)	35	235	235	$(R_b \approx 17.3\Omega)$ 275 $(I_b \approx 2.2$ kA$)$	7.86
Energy (J)	4900	3452	2750 (6 Blumleins)	363 (6 Gaps)	7.4

Table 7-6 Linear Induction Accelerators developed since 1984 and with energy greater than 10 MeV.

Accelerator	Type	Location	Energy (MeV)	Current (kA)	Duration (ns)	Year status
ATA	A	Livermore USA	50	10	50	1984 in operation
Upgraded Radlac II	R	Sandia USA	20	40	40	1988 in operation
LIA-30	R	EPRI Russia	40	100	20	1989 in operation
Radlac II/SMILE	R	Sandia USA	14	100	40	1990 in operation
SABRE	R	Sandia USA	10	250	40	1991 Testing
DARHT	A	Los Alamos USA	16	3	60	1992 Testing

A=Astron type; R=Radlac type.

7-3 Laser Driver

(a) General Remarks

Using intense laser beams is one of the potentially feasible schemes to achieve inertial confinement fusion (ICF). The laser beam for such applications is required to have approximately the following characteristics. Pulse energy greater than 1 MJ, pulse length less than 10 ns and having a beam intensity greater than 100 TW/cm^2. Generation of such a laser beam has been one of the major objectives of the high power pulse research programs around the world. Three types of lasers have been extensively studied in recent years. They are the Neodymium-doped glass (Nd-glass) laser, the CO_2 gas laser and the excimer laser. Though studied quite extensively, at the present time none of these lasers have achieved the capability level required by the ICF application. The key question is how to effectively convert the primary energy of the conventional form into photo energy at extremely high levels of power and intensity. To realize this, one must first achieve efficient population inversion in the lasing medium, i.e. to develop a pumping technique that can achieve optimal amplification of the light intensity.

(b) Pumping Mechanisms

The general idea behind all pumping mechanisms is to convert the conventional energy such as the electric energy stored in the capacitor into atomic or molecular energy of excited state which can be extracted as a laser beam. Various pumping techniques have been developed in the past. For high power pulsed lasers of the category described above, the most suitable pumping methods are the optical pumping, electric discharge pumping and electron beam pumping techniques.

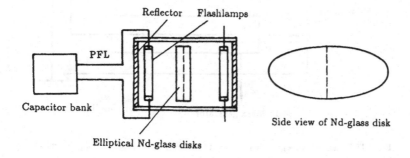

Fig.7-17 Schematic of the cross sectional view of a Nd-glass laser amplifier. The laser beam propagates perpendicular to the plane of the drawing.

In the optical pumping mechanism, population inversion in the lasing medium is achieved by using photon source generated from a flashlamp (e.g. xenon discharge tube) as the exciting source. With such a scheme, a significant number of atoms can be pumped to their upper lasing level before they have a chance to make spontaneous emissions. Upon external stimulation, these atoms transfer simultaneously to their lower state and give rise to the desired laser beam. The technique of optical pumping is the primary mechanism used in lasers with solid or liquid state material, for gaseous state lasers other pumping techniques are generally preferred. Fig.7-17 is a schematic showing the basic structure of the amplifier of a Nd-glass laser. It consists of mainly three parts; a capacitor bank serving as the energy source, a pulse forming line to shape the voltage pulse to the desired form and the laser chamber to accomplish the population inversion process and generate stimulated emissions. The laser beam is extracted in the direction perpendicular to the plane of the drawing. In the laser chamber, there are a number of elliptical Nd-glass disks inclined at the Brewster angle to the beam direction to minimize reflection by the disks. A large number of transverse flashlamps backed by silver-plated crenelated reflectors are arranged at one or both sides of the glass disks and such measures are to provide an optimal amount of light to the glass disks while minimizing absorption by neighboring flashlamps. The flashlamps are usually Xenon gas discharge tubes powered by a set of capacitor banks. As dis-

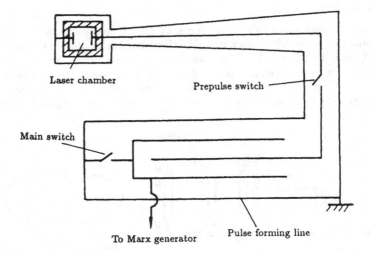

Fig.7-18 Schematic of a gas laser pumped by electric discharge mechanism. The laser beam propagated in the direction perpendicular to the plane of the drawing.

charge of the flashlamp does not require too high a voltage, the capacitor banks in this case are usually charged between 20 to 30 kV.

If an electric discharge is maintained in a gaseous lasing medium such as CO_2 gas, some free electrons will be created due to impact ionization of the gas molecules. Energetic collisions of the electrons and photons with the gas molecules or ions can create excited states. Such excited states can transfer energy to the upper lasing levels by collisions thereby achieving population inversion in the lasing medium. This combination of collision excitation and energy transfer is the basic pumping mechanism used in all electrically discharged gas lasers. However, lasers using such a pumping mechanism cannot readily scale up to a large size due to instability developed during the discharge process. Fig.7-18 is a schematic showing the basic structure of a gas laser pumped by electric discharge mechanism. The essential element of the laser chamber is a pair of parallel plate electrodes connected to the output end of a pulse forming line (a Blumlein in this example). The laser chamber is filled with the lasing gas and the laser beam propagates in a direction perpendicular to the plane of the drawing. The PFL in this case is charged by a Marx generator rather than by a capacitor bank because efficient pumping of the lasing gas under such conditions requires both high voltages and currents with fast rising times. The main advantage of the discharge pumping mechanism is that, unlike in the electron pumping mechanism which shall be discussed in the next section, there is no need to use the anode foil to separate the lasing gas from the pumping source, hence eliminating the problem of heating up the foil. For this reason, discharge pumping mechanism is more suitable for repetitive operations.

Table 7-7 Major parameters and characteristics of performance of the KrF laser shown in Fig.7-19

Electron diode and beam						
Voltage	Current	Energy	Pulse width	Impedance	Cathode area	Anode foil
620 kV	160 kA	8 kJ	80 ns	2.5 Ω	120x75 cm^2	50 μm Al

Laser chamber				Performance		
Volume	Aperture diameter	gases	Pressure	Marx Output	Laser Pulse Energy	Laser Pulse Width
23 L	20 cm	Ar/Kr/F	0.25 MPa	1.1 MV	100 J	100 ns

Reproduced from Ref.[17].

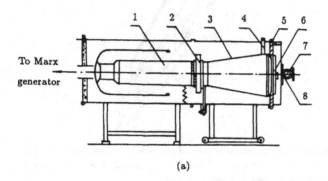

(a)

Fig.7-19(a) Schematic of an *e*-beam pumped excimer laser built by the Institute of Atomic Energy of China. The laser beam propagates in the direction perpendicular to the plane of the drawing [17]. 1. Blumlein transmission line; 2. prepulse switch; 3. tapered transmission line; 4. resistance divider; 5. current shunt; 6. cathode; 7. laser chamber; 8. anode foil.

The third kind of pumping mechanism, also the most attractive one as far as high power gas lasers are concerned, is the electron beam (*e*-beam) pumping mechanism. The attraction is mainly attributed to its suitability of scaling up to a large mode volume and high pulse energy[16]. The basic principle of the *e*-beam pumping technique is similar to that of the discharge pumping mechanism except in this case the electron source is generated externally by some separate device whereas in the latter technique the electrons are created within the lasing gas by impact ionization process. A schematic of the basic structure of the *e*-beam pumped gas laser built by China Institute of Atomic Energy (CIAE)[17] is shown in Fig.7-19(a). This is a KrF laser and the system consists of four basic units:

a Marx generator as the primary energy source, a Blumlein as the PFL, an *e*-beam diode providing the required electron source and finally the laser chamber for the generation of the laser beam. A detailed description of laser chamber is given in Fig.7-19(b). The laser cavity is filled with KrF gas and it is isolated from the vacuum diode by thin foils as indicated in the figure. During operation, the *e*-beam emitted from the cathode enters the laser cavity via these thin foils with negligible attenuation. This laser system is a typical excimer laser pumped by the *e*-beam and belonging to the one hundred joule class. The major parameters and characteristics of performance of this laser reported in Ref.[17] are given in Table 7-7.

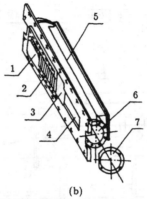

(b)

Fig.7-19(b) Three dimensional view of the laser chamber and anode foil shown in Fig.7-19(a). 1, anode foil; 2, hibachi; 3, main foil; 4, *e*-beam window; 5, laser chamber body; 6, laser window; 7, reflector.

Table 7-8 Optimal Performance of Several Big Lasers

Name	Nova	Antares	LAM
Lasing medium	Nd-glass	CO_2	KrF
Wavelength (μm)	1.06	10.6	248 nm
Pulse Energy (kJ)	300	40	10
Pulse Length (ns)	3	950	650
Efficiency (%)	< 1	5	6.5

This table is compiled basing on available information at the time of writing. Data may not be complete or up-to-date.

(c) Scaling Up Laser Performance

As far as the ICF application is concerned, the performances of the three lasers described in the preceding sections are far below the required level. Even the performances of the bigger lasers are still orders of magnitude below that level. Table 7-8 shows the optimal performance of some big lasers in existence.

From these data, we can see that, in terms of pulse energy the Nd-glass laser is apparently the best. However, its poor efficiency makes it unsuitable for large scale applications. (Here the efficiency is defined as the ratio of the optical energy output to the electrical energy stored). The next best in this respect is the CO_2 laser which has a reasonably high pulse energy and efficiency. The problem with this laser is its long wavelength. It is well known that the long wavelength radiation does not interact with the target material as effectively as the short wavelength radiation does. Therefore CO_2 laser is not a suitable candidate either, as far as ICF application is concerned. The only laser left which appears to be most promising is the KrF laser. However, from Table 7-8, one can see that the present level of overall performance of the KrF laser is still far below the ICF requirements. Developing some means that can significantly scale up the performance of the KrF laser is apparently necessary. In the following section, we shall see how it can be done.

In the rare gas halide (RGH) lasers, the effective gain g is approximately related to the light intensity J by the relationship

$$g = \frac{g_0 J_s}{J_s + J} - \alpha \qquad (7\text{-}2)$$

where g_0 is the small-signal gain, J_s is the saturation light intensity and α is the absorption coefficient of the gas. From Eq.(7-2), we can see that the gain g is inversely proportional to the light intensity J and at certain values of J the gain will become negative. For KrF lasers, it can be shown that the optimum extraction efficiency occurs at $g/\alpha \approx 15$ which gives the maximum attainable intensity $J_m \approx 4$ MW/cm^2. As $J_m = P/A$, that means to achieve high power P, one must use large area A, i.e. large aperture for the laser chamber is necessary.

The other feature of the RGH lasers is that their radiative lifetimes are relatively short (~ 3 ns) so they can be effectively pumped only during a short period of time, otherwise spontaneous emission will occur. However, the pumping pulse of electron beams is relatively long, usually having a pulse length of several hundred ns. To achieve high efficiency it is necessary to inject a maximum amount of pumping energy into the lasing gas during the desired time. This requires some means to multiplexing the short pulses within the time duration of the long pulse. There are other reasons that multiplexing of short pulses is attractive and necessary. In the ICF application, due to power level considerations, pulses in the 5–20 ns range are most appropriate. However, pulsed power facilities such as e-beam diode, operates more efficiently at pulses of several hundred ns duration. Therefore some form of pulses compression is also necessary.

Several methods have been developed in the past to achieve this purpose. The most notable ones are the Raman pulse compression and angular multiplexing methods. The first method involves generating the excimer pulse and then passing it through a Raman-active medium, allowing the nonlinearity of backward stimulated Raman scattering to compress the pulse. In the second method the laser amplifier is operated on a large time scale, usually in the μs to ms range.

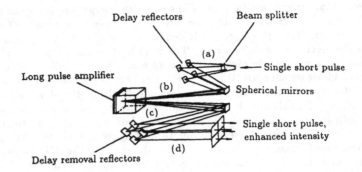

Fig.7-20 Illustration showing the procedure of pulse encoding in angle and time [18].

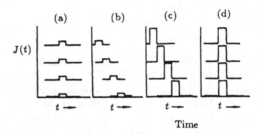

Fig.7-21 Time sequence and pulse shapes corresponding to the illustration given in Fig. 7-20.

During that time a number of short pulses that have been encoded in angle and staggered in time pass through it. After being sequentially amplified, these short pulses are differentially delayed so that they all arrive at the target simultaneously as a single, short pulse of enhanced intensity. The entire procedure of pulse encoding (or multiplexing) in angle and time is shown in Fig.7-20. A short pulse generated from the laser oscillator is sliced into several beams by the slicer arrays (or beamsplitters). For simplicity, only four beams are shown in the figure. The four beams are differentially delayed by the delay reflectors, so the four initially simultaneous pulses arrive at the laser amplifier in sequence, where they make two passes and enhance their amplitude. After they are reflected by the delay removal reflectors, their optical delays are removed and once again they become simultaneous but with enhanced intensity[18]. The corresponding time sequence of the pulse slicing, staggering and amplification is shown in Fig.7-21. In the figure parts labeled with (a), (b), (c), and (d) correspond to the portion labeled with the same notations in Fig.7-20. By employing this method in conjunction with large aperture modules, researchers have been able to achieve significant improvement in the performance of Excimer lasers. In the following section, we shall use the Aurora laser to illustrate it.

(d) Application of Angular Multiplexing

LAM in the Aurora ICF System at LANL, USA is the most energetic Excimer laser in operation to date[16]. Its laser chamber has an extraction volume of 2000 l and with a window aperture of 10^4 cm^2. The lasing gas is a mixture of $Kr/F_2/Ar$ at the pressure range between 600 to 1200 Torr. The gas is pumped by e-beams from two opposite directions at current density 12 A/cm^2. The designed output light energy is 20 kJ. When it is operated as an oscillator, it has produced more than 10 kJ with an intrinsic efficiency of 6.5%. The Aurora system consists of four amplifiers in series configuration with the LAM as its final amplification stage. It produces a pulse train of 5 ns pulses which are optically encoded into 96 separate beams. These 96 separate beams of 5 ns pulse are sent simultaneously to the target for ICF experiments. Fig.7-22 is a block diagram showing the basic set-up and operating principle of the Aurora ICF system at LANL[19].

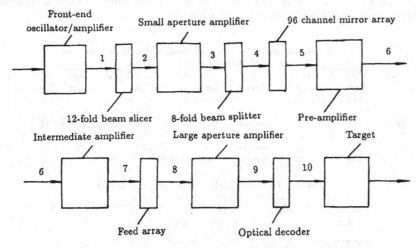

Fig.7-22 Block diagram showing the basic set-up and operating principle of the Aurora ICF system. Explanations for these arrows labeled with number are given in the text.

In the operation, a single laser pulse of 0.25 J energy and 5 ns length is generated from the front-end oscillator/amplifier. This is indicated by arrow 1 in the figure. This pulse is sliced into 12 sequentially separate pulses of the same length but decreased energy as indicated by arrow 2. These 12 pulses are amplified sequentially by the small aperture amplifier (SAM) and form a pulse train of approximately 100 ns length with an energy of 5 J. This is denoted by arrow 3. These 12 pulses are further split and encoded into a pulse train consisting of 96 pulses of 5 ns length with decreased energy as indicated by arrows 4 and 5. At point 5, the energy of these pulses is reduced to 1 J as a result of splitting and encoding. After amplification by the pre-amplifier (PA) these pulses emerge (indicated by arrow 6) as a pulse train of 500 ns length consisting of 96 pulses with an energy of 50 J. Upon further amplification by the intermediate amplifier (IA), the energy of the pulse train enhances to 2 kJ (arrow 7) . Finally the energy of the pulses is beefed up to 20 kJ by the large aperture module (LAM) as indicated by arrow 9. At this stage, the 96 sequentially separate pulses are in a situation qualitatively similar to that shown in Fig.7-21(c). Before they hit the target, the time delays between them must be removed. This is accomplished by the optical decoder which functions in the same way as that of the delay removal reflectors shown in Fig.7-20. Arrow 10 represents the 96 pulses arriving at the target with simultaneity similar to that shown in Fig.7-21(d) . A summary of the nominal design specifications for the original Aurora amplifiers is reproduced from Ref.[16] and presented in Table 7-9. The Aurora ICF system is designed basically for a single-shot operation. Table 7-10 shows the basic parameters of some high power Excimer lasers capable of rep. rate operation.

Table 7-9 Summary of nominal design specifications for original Aurora amplifiers [16]

Unit	PFL Pulse Length (ns)	E-Gun j(in gas) (A/cm²)	E-Gun Area (cm²)	Input Light Energy (J)	Output Light Energy (J)	Stage Gain	Clear Aperture (cm)
SAM	100	12	12x100	0.25	5	20	10×12
PA	650	10	40x300	1	50	50	20×20
IA	650	10	40x300	50	2000	40	40×40
LAM	650	2×12	100x200	2000	10,000 to 20,000	10	100×100

Table 7-10 Some high power Excimer lasers capable of rep. rate operation

Laser name	Medium & Wave-Length	Ave. Power (kW)	Rep. Rate (Hz)	Pulse Energy (kJ)	Builder
EMRLD	XeF, 353nm	4*	100	40	AVCO
Northlight	XeCl, 308nm	4*	50*	100*	NED/TTC
EMRLDO	XeF, 353nm	1.3	100	17	Maxwell
Northlight	XeCl, 308nm	0.7	10	84	NED/TTC
EMRLDO	XeF, 353nm	> 0.35	40		Maxwell

*=potentially capable
NED=Northrup Electronics Div.
TTC=Thermo Electron Technologies Corp.

7-4 Electromagnetic Pulse Simulator

In a nuclear explosion, strong electromagnetic pulse (EMP) with electrical field intensity in the 10^5 V/M range is generated which can cause severe damage to electrical and electronic systems. Fig.7-23 is a list showing the damage thresholds of some electronic components commonly employed in electrical and electronic systems. The threshold power P_f that would cause the electronic component to fail to function may be expressed approximately by[20]

$$P_f = kt^{-1/2} \tag{7-3}$$

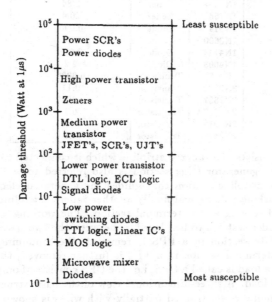

Fig.7-23 EMP damage threshold for some commonly used electronic components.

where t is the time during which the power P lasts, k is called the Wunsch constant and its values for some electronic components are given in Table 7-11. From these data, we can see that most of these electronic components are strongly susceptible to the nuclear electromagnetic pulse (NEMP) effect. Study of the effect is usually done by using a simulator to generate specific EM pulses of the following characteristics: peak electrical field in the 10^5 V/M range, pulse rise time less than 10 ns and falling time in the several hundred ns to 1 μs range. These pulses are used to carry out the desired tests. The basic structure of the EMP simulator consists of essentially three sections: a Marx generator to supply the required power, a specially designed stripline to serve as a PFL as well as an antenna and a terminal load to absorb the energy. One of the early versions of the EMP simulator has been discussed previously in Chapter 4 and its basic structure was shown in Fig.4-31. The newer designs basically fall into three types[21]: the parallel plate simulator (PPS), the triangular plate simulator (TPS) and the conical plate simulator (CPS) as shown respectively in (a),(b) and (c) of Fig.7-24.

Table 7-11 Wunsch Damage Constants for some Transistors and Diodes [20].

Device	Type	K ($W\text{-}sec^{1/2}$)
1N750A	Zener	2.84
1N756	Zener	20.4
1N914	Diode	0.096
1N3600	Diode	0.18
1N4148	Diode	0.011
1N4003	Diode	2.2
2N918	Transistor	0.0086
2N2222	Transistor	0.11
2N2857	Transistor	0.0085
2N2907A	Transistor	0.1
2N3019	Transistor	0.44
2N3440	Transistor	1.1

The PPS consists of a tapered stripline which is connected to the power source such as a Marx generator (this section is also referred to as a wave launcher), a parallel-plate stripline section that guides the generated electromagnetic field towards the working volume and finally another tapered stripline terminated with a resistive load serving as the terminal section. The working volume, where the instruments under test are to be placed, is between the two parallel plates. When the parallel plate section in a PPS is removed, the remaining structure forms a TPS. If the terminal section in a TPS is further removed, then the remaining structure with a load is called a CPS, i.e. the CPS consists of only a wave launcher and a terminal load. A three-dimensional view of the basic structure of a parallel-plate EMP simulator constructed partially with wires is shown in Fig.7-25. The fields generated by such a simulator are not uniform and rather complicated. The fields in the two conical-plate regions are cylindrical waves propagating in the $+y$ or $-y$ direction. In the parallel-plate region, if the elevation angle ϕ is small, then the fields are predominantly transverse electromagnetic waves (TEM mode) and they can be expressed approximately by

$$E_z = \frac{V}{h} \quad \text{and} \quad H_x = \frac{I}{a} \tag{7-4}$$

In Eqs.(7-4), E_z and H_x are the z-component of the electric field and x-component of the magnetic field, respectively, a and h can be identified in Fig.7-24(a), V is the voltage across the parallel plates and I is the current on the plates. All these

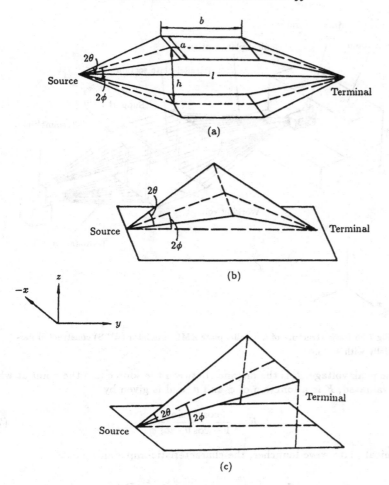

Fig.7-24 Three types of EMP simulator. (a) The parallel-plate simulator (PPS);
(b) The triangular-plate simulator (TPS); (c) The conical-plate simulator (CPS).

quantities are functions of time and space. Under certain conditions, the peak value of the electric field E_p for the wave launcher can be expressed analytically by[22]

$$E_p = \frac{Z_0}{2\pi Z} \frac{V_p}{R} F \qquad (7\text{-}5)$$

where $Z_0 = (\mu_0/\epsilon_0)^{1/2}$, Z is the characteristic impedance of the stripline employed,

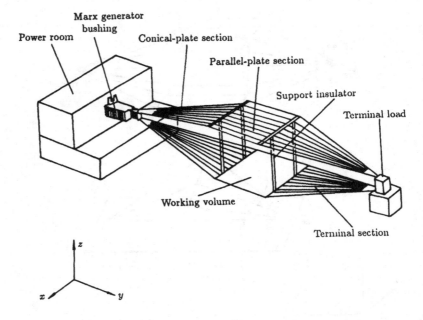

Fig.7-25 Basic structure of a parallel plate EMP simulator (PPS) constructed partially with wires.

V_p is the peak voltage, R is the distance between the source and the point at which E_p is evaluated, F is a function of ϕ and θ, and is given by

$$F = \frac{\cos \theta \cdot \sin \phi}{1 - \cos \theta \cdot \cos \phi} \qquad (7\text{-}6)$$

For conical-plate wave launcher, the characteristic impedance is

$$Z = \frac{Z_0}{2\pi} \int_0^\theta \ln \frac{1 - \cos \omega \cos 2\phi}{1 - \cos \omega} F' d\omega \qquad (7\text{-}7)$$

where $F' = 1/(\pi \sqrt{\theta^2 - \omega^2})$. From these expressions, we can see that the E_p value is dependent on the parameters ϕ and θ, and is independent of the parameters a, b, h and l of the simulator. This implies that the peak fields E_p is solely determined by the wave launcher and the parallel-plate section has no effect on it. This is true only when the voltage pulse generated from the wave launcher has not suffered any reflection yet, because during this period the parallel-plate section serves as a uniform transmission line and transmits whatever coming from the wave launcher without modification. When the reflected pulse from the terminal section arrives, the effect of the parallel-plate section on the peak field E_p will then become noticeable and important.

Table 7-12 Geometrical dimensions and Peak Voltage of three EMP Simulators

Simulator	l (m)	a (m)	b (m)	h (m)	V_p (MV)
ALECS	115	27	15	13	1.3
ARES	189	40	30	40	4
ATLAS 1	410	72	90	105	10

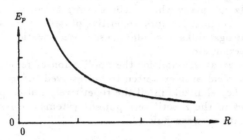

Fig.7-26 Peak value or amplitude E_p as a function of distance R in arbitrary units.

From Eq.(7-5), we can see that, for a conical-plate wave launcher of a given ϕ and θ, the peak value E_p of the electric field is inversely proportional to the distance R. A plot of E_p against R expressed in arbitrary units is shown in Fig.7-26. As the strong field region occurs near the apex where the working volume is small, the usefulness of the conical-plate wave launcher for testing large equipment is rather limited. The main advantage of the conical-plate wave launcher is that its fields propagate outwardly, such property may be necessary for certain applications. If a large working volume is required, such as in the case of testing an aircraft[23], then the parallel-plate simulator shown in Fig.7-25 is appropriate, because the dimension b shown in Fig.7-24 can be made considerably large without much effect on the fields characteristics. The dimensions and peak voltage of some EMP simulators that have been in operation are given in Table 7-12. From these figures, we can see that the overall size of an EMP simulator is significantly larger than a particle beams generator or a linear accelerator of comparable voltage rating. However, from a technical point of view, the major difficulty for building an EMP simulator is not due to its large size but rather it is due to the stringent requirements for the pulse risetime and the tapered stripline. To produce a MV voltage pulse of 10 ns risetime requires sophisticated techniques. The general approach is to employ the "peaking-capacitor technique" discussed previously in Chapter 2, Section 2-3(b). As for meeting the tapered stripline requirements, one may consult the discussion given in Chapter 4, Section 4-4(a) where a rather detailed description of the appropriate approach has been given.

7-5 Repetitive Pulsed Power Generator

In the preceding sections, we have discussed several high power pulse facilities which include the particle beam generators, linear induction accelerators, laser drivers and electromagnetic pulse simulators. All these facilities were designed on the single-pulse basis and are not suitable for high repetition operation. In civilian as well as in military applications, there is a growing demand for generators that can produce high power pulses at relatively high repetition rate. For example, potential applications of electron beams having the following properties: the average power of 0.1 MW or greater, energy of 0.1 MeV or greater and repetition rate higher than 100 Hz, are many which include pumping source for gas lasers, driving source for plasma chemical reactors, repetitive pulsed microwave and x-ray, radiative curing of coatings, links and adhesives, polymerization, e-beam sterilization and sludge disinfection etc.

While there is great demand for the application of repetitively pulsed power systems, some technical problems need to be resolved before one can realize such goals. One of the key elements in all the repetitively pulsed power systems is the switches. Switches in the repetitively pulsed systems are required to recover to their insulating state between pulses.

Table 7-13 General characteristics of several repetitive gas spark gaps.

Voltage (kV)	Rep. rate (Hz)	Specific flow rate ($F \times 10^4$)	Flow pattern	Ref.
220	1000	4	axial	24
100	250	3	axial	25
50	1000	4 to 6	rotary	26

$$F = \frac{\text{gas flow (standard cubic feet per minute)}}{\text{voltage(kV)} \cdot \text{electrode area(cm}^2) \cdot \text{rep} \cdot \text{rate(Hz)}}$$

From Table 5-10, we can see that for high power applications, i.e. both high voltage and current are involved, the most versatile switch to date remains the gas spark gaps. As mentioned in Chapter 5, Section 5(b), for repetitive operation greater than 100 Hz, the time interval between successive pulses is usually less than the natural recovery time for most gases employed, therefore forced recovery must be used—usually by means of forced gas flow. This is one of the key problems that must be satisfactorily resolved before one can successfully employ gas spark gaps for repetitive generation of high power pulses. Table 7-13 gives some pertinent data about the general characteristics of several gas spark gaps employed in repetitive operation. The values of the specific flow rate F given in the table may furnish some useful references for the design of repetitive gas spark gaps. If the application involves only low power level, i.e. either the voltage or the current or both are relatively low, then the use of a gas spark gap may not be necessary, ordinary thyratrons or low pressure gas gaps are sufficient to do the job. The advantage of the thyratron and the low pressure gas gaps is that they do not require forced gas flow under repetitive operations as indicated in Table 5-10. In these cases the repetitive system under consideration involves inductive circuit or the application is for final stage compression of the pulse, then the magnetic switch would be the appropriate choice, as from Table 5-10, we can see that the magnetic switch can work at relatively high power levels without the recovery problem.

In addition to the switch consideration, in a repetitive pulse power system one must also pay special attention to the aspects concerning dielectric material and voltage amplifier. Because most of the dielectric strength data were evaluated under static or single-pulse condition. Under high repetition rate operation, their behavior can be quite different, therefore some modification of the design parameters is necessary. When the single-pulse data are used, a safety factor of 2 to 3 is usually required for the design of a comparable repetitive system. As for the voltage amplifier, most high power pulse systems employ either Marx generators or transformers. The failure rate of a system increases with the number of components it employs. A Marx generator generally uses more components, chiefly gas switches, than a transformer. Furthermore, it requires a large facility to supply forced gas flow for the large number of gas switches associated with it. For this reason and others, in a repetitive high power pulse system, the transformer is more practical and useful than the Marx generator as a voltage amplifier. In the following sections, we shall use two examples to illustrate how these ideas are practically implemented.

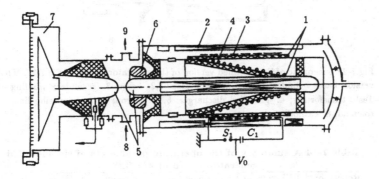

Fig.7-27 Schematic of a repetitive e-beam generator having a rectangular cathode surface [27]. 1–PFL; 2–source housing; 3–primary winding; 4–secondary winding 5–gas spark gap; 6–ring; 7–vacuum diode; 8–gas flow inlet; 9–gas flow outlet; S_1–thyratron switch; C_1– primary source capacitor; V_0–primary charging voltage.

Fig.7-27 is a schematic showing the basic structure of a repetitive high power e-beam generator developed by the Institute of High Current Electronics of the former USSR[27]. The generator was designed to produce a rectangular e-beam of

10×100 cm^2 and with an average power of 5.5 kW. The other operating parameters of the generator are as follows: repetition rate of 100 Hz, diode voltage of 400 kV, current of 8 kA, pulse width of 25 ns, Tesla transformer coupling coefficient=0.6, ratio of secondary voltage to primary voltage=900 kV/22 kV. Upon modification of this generator, the repetition rate has been substantially increased[28]. One of the major modifications carried out was to replace the rectangular e-beam diode with a high impedance diode of cylindrical form (Fig.7-28) so that well focused e-beams could be produced. The second major modification to the generator was to reduce the ratio of the primary winding to the secondary winding to $W_1/W_2 \approx 1/2000$ such that the thyratron switch S_1 can be operated in the voltage range of 500 to 600 volt whereas in the original design it was 22 kV. This measure is essential for higher repetition rate operations. After these modifications, the generator has been capable of operating with a repetition rate of 1 kHz. A summary of the operating parameters of the modified generator is shown in Table 7-14. From the table, one can see that as the repetition rate increases, the corresponding e-beam energy decreases.

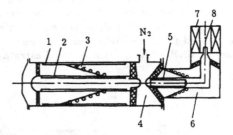

Fig.7-28 Schematic of the modified version of the generator shown in Fig.7-27. This version has a cylindrical vacuum diode [28]. 1,2. cores; 3. secondary winding of Tesla transformer; 4. gas-filled discharger; 5. trigatron; 6. transmission line; 7. solenoid; 8. cathode.

Table 7-14 A summary of the operating parameters of the modified generator shown in Fig.7-28.

Repetition rate (Hz)	PFL voltage (kV)	e-beam energy (keV)	e-beam average power (kW)	Pulse width (ns)
100	1200	800	15	20
500	750	500	25	20
1000	520	350	25	20

The main feature in the design is the conically formed Tesla transformer. Should a Marx generator have been employed as the voltage amplifier, the technical problems faced would be far more complex and difficult. As mentioned before, one of the reasons is that a Marx generator involves a far greater number of components than a comparable transformer does and thus the former is more susceptible to failure. In this design, an oil-filled coaxial line with an electrical length of 10 ns serves as the PFL in which a Tesla transformer is built. The primary winding of the Tesla transformer is placed on the inner surface of the outer conductor of the coaxial line and has a single turn. The secondary winding is wound on a hollow truncated cone and is electrically connected to the inner conductor of the coaxial line as shown in Fig.7-27. The Tesla transformer has a turn ratio of 2×10^3. An equivalent circuit of the generator is shown in Fig.7-29 in which S_2 and C_2 represent respectively the gas spark gap (or the trigatron in Fig.7-28) and the equivalent capacitance of the coaxial line. In operation, the primary capacitor C_1 is charged to a voltage less than 600 volt. Upon closing the thyratron switch S_1, a microsecond voltage pulse is launched onto the coil W_1 meanwhile a voltage pulse of a much higher amplitude is induced in W_2 and charges C_2. When S_2 closes, an ns pulse is launched onto the diode which produces the e-beam pulse specified in Table 7-14. In order to achieve repetitive operations, the switch S_2 was constantly flushed with high pressure N_2 gas. Several S_1 were operated in parallel configuration so that the total current could be shared by a large number of thyratron switches.

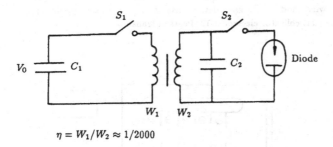

$$\eta = W_1/W_2 \approx 1/2000$$

Fig.7-29 Equivalent circuit of the e-beam generator shown in Fig.7-27 and 7-28. Here S_2 stands for the gas spark gap shown in Fig.7-27 (or the trigatron shown in Fig.7-28) and C_2 is the equivalent capacitance of the coaxial line.

The following is a repetitive pulsed power generator built on a different principle. Fig.7-30 is the cross sectional view of the e-beam generator[29]. This generator is built on the basis of wire ion plasma (WIP) mechanism. Its basic structure consists of mainly a cylindrical cathode, a plane grid and a thin wire anode which is modulated at frequencies of desired magnitude. The entire structure is enclosed in a discharge chamber filled with Helium gas at 10–20 mTorr pressure. The basic operating principle of the generator is schematically shown in Fig.7-31. During

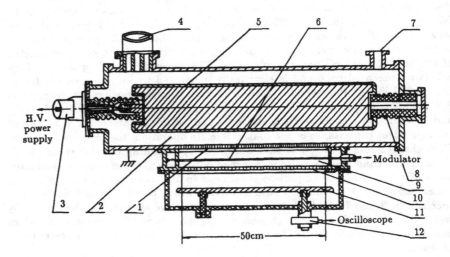

Fig.7-30 Cross sectional view of a wire ion plasma (WIP) e-beam generator. 1, extraction grid; 2, accelerating region; 3, H.V. cable; 4, pump port; 5, cold cathode; 6,thin wire anode; 7,He gas inlet; 8, insulator; 9, discharge chamber; 10, e-beam window; 11, collector electrode; 12, Pearson transformer.

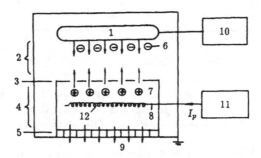

Fig.7-31 Schematic showing the basic principle of the WIP e-beam generator. 1, cold cathode; 2, acceleration region; 3, grid; 4, ions source; 5, foil window and support; 6, electrons; 7, ions; 8, thin wire anode; 9, output electron beam; 10, H.V. DC power supply; 11, modulator; 12, electron orbits.

operation, the thin wire anode is positively modulated at a frequency range of 100 to 1000 Hz. When it is positively charged, some low energy electrons are trapped in the electric field near the thin wire and form helical orbits around the thin wire as indicated in Fig.7-31. These orbits permit the electrons to move along a relatively long path so that ionization of the helium atoms can occur at relatively low pressure environment. Some of the positive ions so produced get through the grid and are accelerated toward the cathode. As a result of ion impact on the cathode surface, some secondary electrons are emitted from the cathode surface. These electrons are accelerated in the cathode-grid region towards the anode. Those electrons gaining sufficient energy will penetrate the chamber window and emerge as an electron beam and the slower ones are trapped at the electric field near the thin wire where they produce more helium ions by impact ionization. In this sense, the grid-anode region is the ion source. When the positive bias on the thin wire is removed, such effects cease to occur. Thus the electron beam is modulated at the same frequency that the thin wire anode is modulated.

The peak voltage applied to the thin wire anode for modulation is between 500–2000 volt at a frequency between 100 to 1000 Hz. The constant d.c. voltage applied to the cathode in this case is about 150 kV. The electron beam current can be estimated by using the following expression

$$I_e \approx \frac{1}{2} \frac{S_g}{S_d} \gamma \eta_g \eta_\omega I_d \qquad (7\text{-}8)$$

In the expression, S_g and S_d are the areas of the grid and discharge chamber, respectively, γ is the Townsend secondary electron emission coefficient, η_g and η_ω are the respective transmission coefficients of the grid and foil window and I_d is the modulation current supplied to the thin wire anode. The discharge chamber employed in this case has the dimensions of $7 \times 4 \times 50$ cm^3 and the helium pressure in the chamber is between 10 to 20 mTorr. The anode is made from 2–6 Tungsten wires of 0.24 mm diameter. The transmission coefficients of the grid and window foil are $\eta_g = 80\%$ and $\eta_\omega = 90\%$, respectively. Some typical results obtained from the generator under these conditions are given below.

e-beam energy:	130-150 KeV
e-beam current density:	250 mA/cm^2
e-beam total current:	63 A
e-beam cross section:	5×50 cm^2
pulse duration:	1 μs
repetition rate:	100 Hz
average output power:	0.94 kW.

Typical current waveforms obtained from the WIP *e*-beam generator are shown in Fig.7-32.

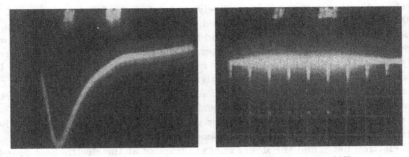

(a) single pulse   (b) repeated pulse at 100Hz

Fig.7-32 Typical current waveforms obtained from the WIP e-beam generator [29].

7-6 Electromagnetic Launcher

(a) General Remarks

The electromagnetic launcher is a device designed to have the capability of launching a projectile with velocity in the kM/s range. The device has a number of potential applications both in civilian and military areas. For example, it may be used to produce hypervelocity particles for impact studies, to develop kinetic weapons or to serve as a cost-saving space launcher. Its basic principle is rather similar to that of a cannon, when the barrel of the cannon is replaced with a pair of rails and the chemical explosive is substituted by an energy source of an electromagnetic nature. Based on the operating principle, electromagnetic launchers may be loosely divided into three basic types. They are the rail gun, induction coil gun and electrothermal gun. Each of these guns have attractive features as well as limitations. In the following, we shall discuss each of them respectively.

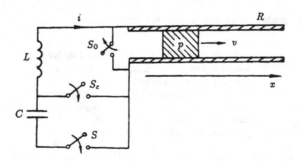

Fig.7-33 A rail gun powered by a capacitively charged inductor. S is the starting switch; S_c is a closing switch S_0 is an open switch.

(b) Rail Gun

Fig.7-33 shows the basic structure and equivalent circuit of a rail gun powered by a capacitively charged inductor. It consists of mainly a primary energy source C, an energy storage L, a pair of parallel rails R and the projectile P to be launched. Initially the open switch S_0 is closed. When the starting switch S closes, a current will flow and the inductor L is charged. At a proper time, when switch S_0 opens and switch S_c closes, the current will flow through the projectile P. Interaction between the magnetic field within the closed circuit and the current carried by the projectile produces a Lorentz force which accelerates the projectile along the rails. If frictional force is negligible, the net force acting on the projectile can be expressed by

$$F = \frac{1}{2}i^2\frac{dL}{dx} \tag{7-9}$$

where i is the current going through the projectile, dL/dx is the inductance gradient along the rail and expressed in Henry per meter. The terminal velocity of the projectile at the exit is then

$$v = v_0 + \frac{1}{2m}\frac{dL}{dx}\int_0^t i^2\,dt \tag{7-10}$$

where v_0 is the injection velocity and usually is taken to be zero, m is the mass of the projectile. dL/dx value is typically in the order of $\mu h/M$. From Eq.(7-10), we can see that for a given mass m, the velocity is proportional to $\int_0^t i^2\,dt$. That implies that a high current and a long pulse duration are required to achieve high velocity. In all the previous discussions we have nearly in every case stressed that achieving high power is the main objective. In the present case, however, it is not so. As one can see from Eq.(7-10) , the terminal velocity is mainly determined by the current intensity i and pulse duration t and is independent of the voltage, hence power is not the main concern as long as one can achieve high current intensity. Alternatively, one may look at the situation in another way. The terminal velocity of the projectile can be also expressed by the relationship

$$v = \sqrt{2\epsilon/m} \tag{7-11}$$

where ϵ is the total kinetic energy of the projectile. It is obvious that unless the primary energy source can supply sufficient energy, the projectile can never attain the desired velocity. For this reason, development of a large energy source has been the chief task in the research of rail guns.

The energy sources that are scalable, hence potentially suitable for rail gun application include capacitor banks, inductors, batteries and inertial machines such as the Homopolar generator. Investigations of these energy sources for rail gun application have been carried out by many institutions around the world and some significant progress has been made in recent years. One of the examples (Theoretically this example should be treated as an induction coil gun. Here, the main concern is for the energy source) is the CEM-UT (Center for Electromechanics at the University of Texas) program in which they have employed homopolar generator (HPG) of 60 MJ capacity as the primary energy source[30]. The HPG is capable of delivering 9 MA current at 100 V voltage. The rail gun consists of a 10 m long, 9 cm round-bore launcher powered by six HPG charged inductors[31]. The circuit is basically similar to that shown in Fig.7-33 when the capacitor C is replaced by the HPG. The armature employed is hybrid type, made from Al of 1.1 kg mass. The total mass of the package is 2.44 kg. The terminal velocity attained is 2.6 kM/s. The experiment was carried out in a vertical test range of 50 m deep. Basing on these data, the total kinetic energy of the projectile in this case is about 8.3 MJ.

Another example is the Battery Power Supply (BPS) program of the US Air Force[32]. In this system, a large number of automotive batteries are employed to supply the primary energy for the rail gun. A scheme of using 22880 batteries to deliver a current of 2.5 MA to a 200 MJ inductor has been planned. A 2.15 MA current has been actually observed via a system shown in Fig.7-34. The BPS system shown in this block diagram consists of a total number of 13728 batteries and each of them is capable of delivering a maximum current of 1.8 kA. These batteries are divided into 858 strings and each string consists of 16 (or 8) batteries in series connection. The 858 strings are further grouped into 36 gangs and each gang consists of 24 strings in parallel configuration. During operation, the switch S_0 is closed initially. When the inductor L is fully charged by the batteries, S_0 opens thereby transferring the stored magnetic energy into kinetic energy of the projectile.

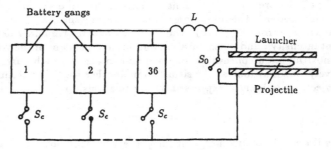

Fig.7-34, 36-Gang Battery power supply (BPS) system for Rail Gun (1) 1 gang consists of 24 strings in parallel connection; (2) 1 string consists of 16 automotive batteries in series. (3) the peak output current of a single battery is 1.8 kA. (4) L=200 MJ inductor, S_0=open switch; S_c=contactor switch.

(c) Induction Coil Gun and Electrothermal Gun

The basic feature of the induction coil gun is that the propulsive force acting on the projectile is arising from the interaction between a moving coil and the stationary ones. The force is attractive if the currents flow in the same direction in both the moving and stationary coils, and repulsive when the currents flow in opposite directions. Depending on the type of armature employed, induction coil guns can be built in three different forms: the wound armature gun[33], the solid armature gun[34] and linear induction gun[35]. In this section we shall illustrate the basic principle of the most simplest form only.

Fig.7-35 shows the basic structure of the induction coil gun in its simplest form[36]. It consists of a pair of feed rails connected to some dc source. There is a longitudinal array of stationary coils along the rails to form a gun barrel. These stationary coils are usually powered by a capacitor bank and sequentially, properly switched on and off according to the position of the moving coils. Attached to the moving cylinder are several smaller coils which draw current from the dc source via the commutators (or brush) and the feed rails. If the current direction in the stationary coils are properly adjusted and synchronized, the moving coils shall be attracted to the stationary coils in front of them and repelled by those coils behind. In this way the moving coils are accelerated continuously along the feed rails. The force acting on the moving coils can be written as

$$F = i_1 i_2 \frac{\partial M}{\partial x} \qquad (7\text{-}12)$$

where i_1 and i_2 are respectively the currents in the moving and stationary coils, $\partial M/\partial x$ is the mutual inductance gradient due to the motion in the $-x$-direction of the moving coils.

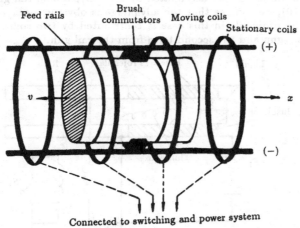

Fig.7-35 Schematic showing the basic principle of the induction coil gun [36].

In the solid armature induction coil gun, there is no direct contact between the armature and the gun barrel. The gun barrel is formed with a series of coils surrounding a tube-shaped flyway. The projectile rides in the flyway with a small clearance. By using this type of coil gun, SNL in USA has achieved a projectile velocity of 335 M/s for a projectile of 5 kg mass[37].

The basic process involved in both the rail gun and induction coil gun is essentially a conversion of the electromagnetic energy supplied by the energy source into the kinetic energy of the projectile. In an electrothermal gun, however, the energy conversion process is somewhat different. The electromagnetic energy from the source in this case is first converted into thermal energy from which it is converted into kinetic energy of the projectile. In an electrothermal gun such conversion processes can be accomplished in various ways depending on the type of designs employed. Fig.7-36 is a schematic showing the basic structure of a hybrid electrothermal rail gun designed on the basis of a plasma pulse accelerator (PPA)[38]. It consists of an end-cup anode joined with an insulating tube that forms the gun body. Behind the projectile there is a discharge chamber filled with an insulating material which could be gaseous, liquid or solid. The end-cup anode and the projectile is electrically connected with a fuse and the front end of the projectile is electrically connected to the grounded rails by a piece of thin wire. When switch S_1 closes, intense current causes the fuse to explode and an electric arc nearly at the meantime is initiated in the liquid. Intense heat created from these processes quickly evaporates the liquid thus producing a high pressure region behind the projectile. Pressure generated by such means can be more than 10^3 bar which may be translated into projectile velocity in the kM/s range. If S_2 closes properly at the time when the projectile just passes the end of the insulating tube, the projectile will be further accelerated by the conventional rail gun mechanism. From Eq.(7-10), we can see that the advantage is obvious, because the terminal velocity of the projectile in this case is contributed by two sources v_0 from the electrothermal gun and the second term from the rail gun.

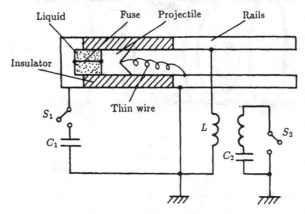

Fig.7-36 Schematic showing the basic structure of an electrothermal-rail gun hybrid.

References of Chapter 7

[1] T. A. Mehlhorm et al., *9th Int. Corf. on High Power Particle Beams*, 1, (1992), 31.

[2] J. D. Shipman Jr, *4th Symp. on Eng. Problem of Fusion Research*, Washington D. C., (1971).

[3] Wang Ganchang, *High Power Laser and Particle Beams* (in Chinese), 1, (1989), 1.

[4] G. W. Forster et al., *J. Vac. Sci. Tech.* 12, (1975), 1177.

[5] G. L. Johnson et al., *IEEE Trans. Plasma Sci,* **PS-8**, (1980), 204.

[6] Qiu Aici et al., *High Power Laser and Particle Beams* (in Chinese), 3, (1991), 340.

[7] S. J. Sackett, Lawrance Livermore Lab. Report UC ID-17814, June, (1978).

[8] B. Goplen et al., User's manual for MAGIC by MRC, Report MRC/WDC-R-068, (1983).

[9] E. L.Neau et al., *5th IEEE Pulsed Power Conf.*, (1985), 772.

[10] J. Delvaux and N. Camarcat, *Proc. NATO Advanced Study Institute on Fast Electrical and Optical Diagnostic Principle and Techniques*, Pascoli, (1983).

[11] N. C. Christofilos et al., *Rev. Sci. Instrum*, **35**, (1964), 886.

[12] A. I. Pavlovskii et al., *Sov. Phys. Dokl.* **25**, (1980), 120.

[13] M. Friedman, *Appl. Phys. Lett.* 41, (1982), 419.

[14] Zhang Enguan et al., *High Power Laser and Particle Beams (in Chinese)*, 5, (1993), 61.
Cheng Nianan et al., *High Power Laser and Particle Beams (in Chinese)*, 4, (1992), 325.

[15] J. J. Ramirez et al., *5th IEEE Pulsed Power Conf.*, (1985), 143.

[16] V. N. Smiley, *SPIE* **1225**, *High Power Gas Lasers*, (1990), 2.

[17] Shan Yusheng et al., *High Power Laser and Particle Beams* (in Chinese), 5, (1993), 5.

[18] R. A. Haas et al., *ICF 80 Topical meeting on ICF Technical Digest*, Optical Soc. of America, (1980), 47.

[19] T. P. Turner et al., *SPIE* **1225**, *High Power Gas Lasers*, (1990), 23.

[20] G. T. Brown, *ITEM*, R & B Enterprises, (1980).

[21] J. J. A. Klaasen, *IEEE Trans. Electromag. Compat.*, **35**, (1993), 329.

[22] H. M. Shen et al., *IEEE Trans. Electromag. Compat.*, **EMC-29**, (1987), 32.

[23] P. L. Rustan Jr., *IEEE Trans. Electromag. Compat.*, **EMC-29**, (1987), 49.

[24] M. Buttram., *4th IEEE Pulsed Power Conf.*, (1983), 361.

[25] A. Ramrus, *IEEE Trans on Elect. Dev.* **ED-26**, No.10, (1979).

[26] J. T. Naff et al., *14th Power Modulator Symp.*, (1980), 21.

[27] F. Ya Zagulov et al., *Prib. Tekh. Eksp.*, No.5, (1976), 18.

[28] N. M. Bykov et al., *Prib. Tekh. Eksp.*, No.2, (1991). 38.

[29] Qi Zhang et al., *9th Int. Conf. on High Power Particle Beams*, 1, (1992), 676.

[30] R. J. Hayes et al., *8th IEEE Pulsed Power Conf.*, (1991), 50.

[31] R. J. Hayes and R.C. Zowarka, *IEEE Trans on Magnetics*, 27, No.1, (1991).

[32] J. B. Cornette et al., *7th IEEE Pulsed Power Conf.*, (1989), 131.

[33] M. W. Ingram et al., *IEEE Trans on Magnetics*, 27, (1991), 591.

[34] R. J. Kaye et al., *IEEE Trans on Magnetics*, 27, (1991), 596.

[35] Z. Zabar et al., *IEEE Trans on Magnetics*, 25, (1989), 627.

[36] H. Kolm and P. Mongeau, *IEEE Spectrum*, April issue, (1982), 30.

[37] B. Henderson, *Aviation Week and Space Tech.*, May 7 issue, (1990), 88.

[38] H. W. Fien et al., *8th IEEE Pulsed Power Conf.*, (1991), 760.

CHAPTER 8
DIAGNOSTICS

8-1 General Remarks

In the preceding chapters we have discussed how to design and operate the various components and systems for high power pulse applications. However, without proper means to measure the quantities yielded from these systems, one can hardly utilize them to generate the desired results. In this chapter we shall discuss the various measuring techniques and devices with which one can obtain reasonably accurate values for the physical quantities involved.

In high power pulse systems, the most common quantities to be measured are the electrical voltage, current, power and energy. Because of the large magnitude and fast temporal behavior of these quantities, ordinary measuring methods are no longer adequate to obtain accurate measurements. One needs to apply special techniques and use special devices. In addition to the quantities mentioned above, measurements of many other quantities are also frequently needed, Table 8-1 lists some of these additional quantities and their measuring methods. In this chapter, however, we shall discuss the details for only those quantities mentioned above. For details of the rest, one is suggested to consult the appropriate references given in Table 8-1.

Table 8-1 Relevant Quantities and their Measuring Methods

Measurement	Method	Ref.
e-beam Current	Faraday Cup	1
Ion Beam Current	Faraday Cup	1
	Biased Charge Collector	2
X-ray Image	X-Pinhole Camera	2
X-ray Dose	Thermoluminescent Dosimeter (TLD)	2
x-ray Waveform and Intensity	Compton Diode (CD) PIN Detector	2
Ion Beam Trajectory	Shodow Box and CR-39 Recording plastic	3
Ion Species and Energy Distribution	Thomson Parabola Ion Energy Analyzer	2
Ion Velocity	Time-of-Flight Spectrometer	4
Ion Spot Size	Ion Pinhole Camera	5

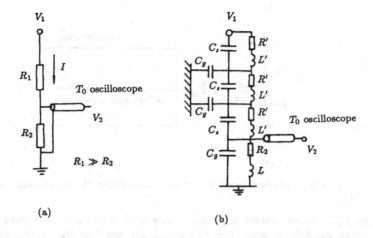

Fig.8-1 (a) Idealized resistive voltage divider; (b) Equivalent circuit of a real resistive voltage divider.

8-2 Voltage Measurement

(a) Resistive Voltage Divider

Owing to its structural simplicity and wide-range applicability, the resistive voltage divider is one of the most widely used devices for voltage measurements. The basic structure of an ideal voltage divider is shown in Fig.8-1(a). It consists of two resistors R_1 and R_2 which respectively serve as the high voltage arm and low voltage one. The low voltage V_2 across the resistor R_2 is directly measured and the unknown voltage V_1 is determined by using the given values of R_1, R_2 and V_2 in conjunction with the formula

$$V_2 = \frac{R_2}{R_1 + R_2} V_1 \qquad (8\text{-}1)$$

This description is only for an ideal situation. In actual cases, circuit inductance and stray capacitance cannot be totally ignored. The equivalent circuit of a real resistive voltage divider should be approximately represented by that shown in Fig.8-1(b) in which the circuit inductance L' and the stray capacitances C_s, C_g have been included. Analysis of a circuit as such is rather difficult. For simplicity, let's first assume that the stray capacitances are negligible, the circuit can then be represented by that shown in Fig.8-2(a). The relationship between the measured voltage V_2 and the unknown voltage V_1 in this case is

$$V_2 = \frac{V_1 R_2}{R_1 + R_2} \left\{ 1 - e^{-\frac{R_1 + R_2}{L_g} t} \right\} \qquad (8\text{-}2)$$

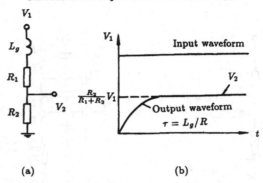

Fig.8-2 (a) Equivalent circuit with inductance included; (b) Actual output waveform.

where L_g is the equivalent total inductance of the circuit. A plot from Eq.(8-2) is shown in Fig.8-2(b). From this figure, one can see that the greater the value of $L_g/(R_1 + R_2)$, the larger the distortion to the original wave-form of the unknown voltage V_1. Therefore it is essential to make the circuit inductance L_g as low as possible. For this reason, the high voltage arm of a divider is usually constructed with liquid or film resistors. The basic rule for designing a resistive voltage divider is that the value of $L_g/(R_1 + R_2)$ satisfies the relation

$$\left(\frac{L_g}{R_1 + R_2}\right) < \frac{t_r}{20} \qquad (8\text{-}3)$$

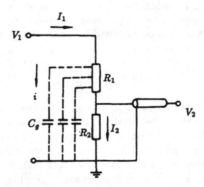

Fig.8-3 An actual divider with current i going through the stray capacitance.

where t_r is the rise time of the unknown voltage to be determined. If the circuit inductance is negligible and only the effect of stray capacitance is considered, then the circuit can be approximated as that shown in Fig.8-3. In this case, the measured voltage V_2 is related to the unknown voltage V_1 by[6]

$$V_2 = \frac{V_1 R_2}{R_1 + R_2} \left\{ 1 + 2 \sum_{n=1}^{\infty} (-1)^n e^{-\frac{n^2 \pi^2 t}{c(R_1 + R_2)}} \right\} \qquad (8\text{-}4)$$

where C is the total stray capacitance between the divider and the ground plane. From this expression, it is clear that only when $C = 0$, is V_2 expressible by Eq.(8-1). In practice, there is always some stray capacitance present in the circuit, therefore the rise time of the measured voltage V_2 is always greater than that of the voltage to be determined. How much greater is determined by the value of $C(R_1 + R_2)$. The general rule for design in this case is

$$C(R_1 + R_2) < \frac{t_r}{0.23} \qquad (8\text{-}5)$$

where t_r is again the rise time of the voltage to be determined. Combining this rule with that given in Eq.(8-3), gives the general rule for practical design of resistive voltage dividers. One should try to follow it as closely as possible.

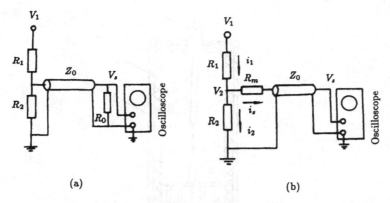

Fig.8-4 (a) A matched resistive voltage divider with parallel connection; (b) A matched resistive voltage divider with series connection.

The other source of measuring error that may result is due to impedance mismatch at the input and output ends of the connecting conductor. For instance, when an oscilloscope is employed to display the output voltage V_2 from the divider, a coaxial cable is usually used to connect the output end of the divider to the input end of the scope. The cable impedance usually is between 50 to 100Ω, its capacitance per unit length is about 57-115 μF/m, and the cable length is usually in the range of a few to a few tens of meters. If the impedances at the two ends of the cable are not properly matched, from Section 4-1(b), we know that the voltage pulse will be reflected at the terminals. Superposition of the original with the reflected pulses may seriously distort the true pulse form. The remedy to this problem is usually to either connect a parallel resistor at the input end of the measuring device or a series resistor at the output end of the divider as shown respectively in Fig.8-4(a) and (b). In Fig.8-4(a) the value of R_0 should be identical to the characteristic impedance Z_0 of the cable and in Fig.8-4(b) R_m is required to satisfy $R_2 + R_m = Z_0$. For $R_m \geq Z_0$ cases, the series connection is not applicable. When the parallel connection is employed, the relationship between the measured voltage V_s and the unknown voltage V_1 is

$$V_s = \frac{V_1 R_2 R_0}{R_1 R_2 + R_1 R_0 + R_2 R_0} \tag{8-6}$$

When the series connection is used, the corresponding relation between V_s and V_1 is

$$V_s = \frac{R_2}{R_1 + R_2} V_1 \tag{8-7}$$

In most practical cases, impedance match of parallel connection is preferable because it is simple to make and free from restrictions.

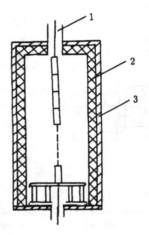

Fig.8-5 Basic structure of a coaxial resistive voltage divider, (1) Coaxial cable; (2) Insulation tube; (3) External shielding [6].

Fig.8-5 shows one common design of the resistive voltage divider in which the high voltage arm is constructed with a series of liquid or film resistors and the low voltage arm is made up of a number of film resistors connected in parallel configuration. In order to compensate for the effect of stray capacitance formed between the low voltage arm and the ground plane, some investigators[7] have designed a voltage divider with an inductor connected in series to the low voltage arm of the divider. By properly selecting a value for the inductor, the response time of the voltage divider has been reduced from 8 ns to 2.8 ns while the peak value of the measured voltage stays uneffected.

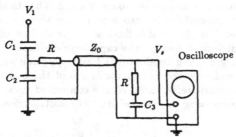

Fig.8-6 A capacitive voltage divider and its connections.

(b) Capacitive Voltage Divider

Capacitive voltage divider has also been widely used for measuring pulsed voltages. The basic structure of the capacitive voltage divider is quite similar to that of the resistive voltage divider except in this case the resistors are replaced with appropriate capacitors. Fig.8-6 shows the circuit of a typical capacitive voltage divider connected to a measuring device through a coaxial cable. Capacitor C_1 and C_2 represent respectively the high voltage and low voltage arms. As C_1 is usually much smaller than C_2, the total capacitance of the divider is $C \approx C_1$ and C_1 normally is in the order of a few pf. Such a small capacitance draws very little current from the main circuit hence the divider has practically no effect on the latter. The other feature of the capacitive voltage divider is its good ability in responding and its bandwidth can be extended up to 1.5 GHz. The main disadvantage is that it is susceptible to the effect of undesirable oscillations present in the output voltage, when there is some inductance in the circuit. One way to lessen the effect is to employ capacitors having the least inductance. The other potential problem of the capacitive voltage divider is the so-called RC coupling effect which causes the output voltage to decay exponentially. To prevent this from happening, the measuring circuit should be properly arranged as shown in Fig.8-6 in which the components are selected to satisfy the relations $R = Z_0$ and $C_1 + C_2 = C_3 + C_k$. In the above relations, Z_0 is the cable impedance and C_k the stray capacitance of the circuit. When these relations are satisfied, the measured voltage V_s can be expressed by

$$V_s = \frac{V_1 C_1}{2(C_1 + C_2)} \tag{8-8}$$

Capacitive voltage dividers are frequently employed in liquid dielectric transmission lines of large scale. Measurement of pulsed powers in the order of 10^{12}W with pulse lengths of the order 10–100 ns requires a suitable voltage divider which must be immune to the strong E-M interference fields produced in the system while keeping the conditions of the system unaffected by the presence of the voltage divider. Fig.8-7(a) shows the basic structure of such a voltage divider[8]. The equivalent circuit of the divider is shown in Fig.8-7(b). In the design, a cylindrical plug is mounted concentrically into a recess on the grounded conductor of the transmission line. The plug end is flush with the surface of the grounded conductor. The liquid dielectric also fills the capacitive divider. The gap spacing between the plug and the grounded conductor is t and opposing area A. The distance between the plug and powered conductor is d and the end area B. R_1 and R_2 in Fig.8-7(b) are dielectric resistances, R_s is a connected resistor for reducing dividers sensitivity and increasing the bandwidth. The voltage division in this case is

$$V_1 = V_2 \frac{Ad}{Bt} \qquad (8\text{-}9)$$

The advantage of this design is that the measurements are independent of effects due to temperature, frequency and impurity in the liquid.

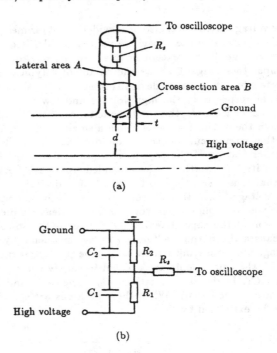

(a)

(b)

Fig.8-7 (a) Basic structure of the capacitive voltage divider attached to the T-L; (b) Equivalent circuit.

(c) Optical Voltage Monitor

By using the linear electro-optic effect (Pockels effect) in certain crystals, voltage can be measured optically. The advantages of this technique are its electrical isolation, immunity to E-M radiation interference, high sensitivity and small size. When capacitive coupling is used, a high voltage of up to 1 MV can be measured. Fig.8-8 shows the electrical and optical arrangement of an optical voltage monitor system employed by Herts and Thomson[9]. The high voltage to be measured is V_0 and the high voltage conductor is capacitively coupled to the Pockels cell (in this case it is a KD*P crystal) via an antenna plate A. The sensor unit is passive and is connected to the laser-detector unit by optic fibers, therefore the laser-detector unit is totally isolated from the HV source. If the Pockels cell is placed between crossed polarizers and a $\lambda/4$ plate, the relation between the input intensity J_0 and output intensity J of the laser light is

$$J = J_0 \sin^2 \frac{\pi}{2} \left(\frac{V}{\beta} + \frac{1}{2} \right) \tag{8-10}$$

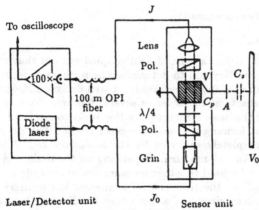

Fig.8-8 Electrical and optical arrangement of an optical voltage monitor system [9].

where $\beta = \lambda/(2n_0^3 \gamma_{63})$ is a constant and is determined by the wavelength λ, optical constants n_0 and γ_{63} of the crystal. V is related to the high voltage V_0 by

$$V = V_0 \left(\frac{C_s}{C_s + C_p} \right) \tag{8-11}$$

where C_p and C_s are respectively the capacitances of the Pockels cell and that between the HV conductor and antenna plate A. By monitoring the output light intensity J and using Eq.(8-10), one can obtain V. Further from Eq.(8-11), the

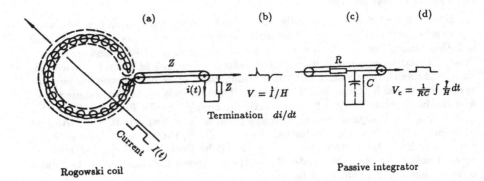

Fig.8-9 Schematic representation of Rogowski Coil and its accessory.

high voltage V_0 can be determined. The main disadvantage of this technique is that the capacitive division has to be calibrated for each different situation.

8-3 Current Measurement

(a) Rogowski Coil

The Rogowski coil is a type of field coupled sensors that utilizes the induced voltage in the secondary coil to determine the primary current. It converts large current into small one, then measures it. A well designed Rogowski coil is capable of measuring currents of the order of several hundred kA and a few ns risetime. Fig.8-9(a) shows the basic structure of a Rogowski coil in flexible form. It consists of N small loops forming a nearly complete circle. The primary current to be determined is completely encircled by the secondary coil. This is equivalent to having a primary coil of one turn. Depending on the design and applications, the Rogowski coil can be used to either measure the magnitude and pulse form (if an integrator is used) or the time rate of change of the primary current. Fig.8-10 shows the equivalent circuit of a secondary winding of the Rogowski coil. The governing equation for the voltage drops along the circuit is

$$\frac{L}{N}\frac{dI}{dt} - L\frac{di}{dt} - (R + Z)i = 0 \qquad (8\text{-}12)$$

where I and i are respectively the primary and secondary currents, L, N, R are respectively the inductance, number of turns and resistance of the Rogowski coil, and Z is the signal resistance. The first term in the equation represents the induced voltage by the primary current $I(t)$. There are three basic ways to apply the Rogowski coil for the measurements of current I.

(I) If the characteristic time t_c, e.g. the rise time of the primary current is such

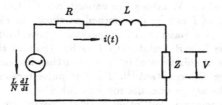

Fig.8-10 Equivalent circuit of a Rogowski coil with integrator. $\frac{L}{N}\frac{dI}{dt}$–Induced EMF in the secondary coil; R–Resistance of coil; Z–Metering resistor; $i(t)$–Current in the secondary coil; V–Voltage drop across the metering resistor.

that

$$t_c \ll \frac{L}{R+Z} \tag{8-13}$$

then the third term in Eq.(8-12) can be neglected, thus we have

$$V = iZ = \frac{I}{N}Z \tag{8-14}$$

By measuring the voltage V across the resistance Z, one can determine I. From relations (8-13) and (8-14), one can see that this type of design is capable of measuring the current pulses of fast rise time and the large magnitude if the value of N is kept large and Z small.

(II) If

$$t_c \gg \frac{L}{R+Z} \tag{8-15}$$

then the second term in Eq.(8-12) is negligible. Accordingly we have

$$V = iZ = \frac{LZ}{N(R+Z)}\frac{dI}{dt} \tag{8-16}$$

This equation shows that the measured voltage V is proportional to dI/dt as indicated in Fig.8-9(b). When the Rogowski coil is employed in this fashion, the device is commonly referred to as an I-Dot probe[10].

(III) If, in the circuit shown in Fig.8-10, a capacitor C is added in series with the signal resistor Z, then the governing equation becomes

$$\frac{L}{N}\frac{dI}{dt} - L\frac{di}{dt} - (R+Z)i - \frac{1}{C}\int i\,dt = 0 \tag{8-17}$$

Under the conditions of $t_c \gg L/(R+Z)$ and C is sufficiently large, the second term and fourth term in Eq.(8-17) can be neglected, then the voltage V_c across the capacitor C can be expressed as

$$V_c = \frac{1}{C}\int i\,dt = \frac{LI}{NC(R+Z)} \tag{8-18}$$

By measuring the voltage V_c across the capacitance C, from Eq.(8-18) the pulse form and magnitude of I can be determined as shown in Fig.8-9(d). The device shown in Fig.8-9(c) is a passive RC integrator whose function is to make the output voltage to be linearly related to I rather than to the time derivative of I. The disadvantage of this type of integrator is that its time response is usually slow due to the large RC constant[11]. For fast pulse measurements, other types of integrators e.g. the active one, are more suitable.

In designing a Rogowski Coil, other than the basic principle discussed previously which should be followed, certain practical aspects must also be properly treated. These include the aspects of shielding the coil from external interference, and leaving a proper slit opening in the shield so that the magnetic field can penetrate into the coil. The basic structure of a practical Rogowski coil is shown in Fig.8-11.

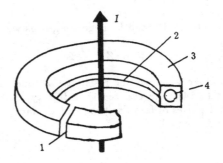

Fig.8-11 Basic structure of a practical Rogowski coil. 1,2–Slit; 3–Metallic shielding; 4–Coil.

(b) Current Viewing Resistor

Current viewing resistor (CVR) is operated on the basis of Ohm's law, i.e. the measurement of the current relies on the voltage developed across certain resistive material when the circuit current flows. The schematic of a tubular CVR is shown in Fig.8-12. On the outer conductor there is an (or more) annular cavity filled with resistive material which is covered with thin foil. When the circuit current I flows along the conductor a voltage drop across the resistive material is formed. By monitoring this voltage, the circuit current I can be determined. If the annular resistive material has an inner radius r, an outer radius a and a length l, then the output voltage V of such a CVR to a step rise input of magnitude I is approximately[12],

$$V = \frac{I\eta l}{\pi^{3/2} a \delta \tau^{1/2}} e^{-\frac{1}{4\tau}} \tag{8-19}$$

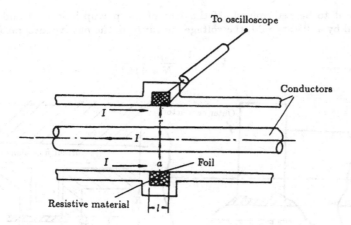

Fig.8-12 Schematic of a tubular current viewing resistor.

Where $\delta = a - r$, $\tau = \eta t/(\mu \delta^2)$, η and μ are respectively the resistivity and permeability of the resistive material and t is time. From Fig.8-12, one can see that the CVR is usually a four terminal network so that the metering contacts are independent of the current-carrying contacts. Even so, proper caution should be observed because its direct connection to apparatus operating at high voltages and carrying a large current may introduce electric interference problem. When CVR is constructed in a form, i.e. stripline CVR, it can be directly mounted in the current return of systems having plate geometry, such as the laser head of rectangular shape.

(c) Cavity Current Monitor

The basic structure of the cavity current monitor is in some aspects rather similar to that of a CVR. In the CVR the re-entrant cavities which are machined in the current carrying conductors are filled with resistive material whereas in the cavity current monitor they are kept empty. The main difference is that the former is operated on the basis of Ohm's law and the latter is operated on Faraday's law. In the cavity current monitor there is a resistor connected across the cavity as shown in Fig.8-13(a). The connecting lead of the resistor along with the cavity walls form a closed loop which is used as a magnetic pickup coil as illustrated in Fig.8-13(b). When the current flows in the conductor, the re-entrant cavity develops a voltage drop across itself, corresponding to the rate of change of magnetic flux in the cavity. An array of magnetic pickup loops is usually used to sample the flux change in the cavity at different locations along the current path. The signals generated by these pickup loops are combined using cables of the same length at a summing circuit as shown in Fig.8-14. The combined output is then integrated using a passive RC integrator to yield a signal proportional to the current to be determined. If the area of the pickup loop is A and they are separated by a distance Δl, the voltage output from the passive integrator is given by[13]

$$V = \frac{1}{RC} \frac{\mu_0 A I}{(N+1)\Delta l} \tag{8-20}$$

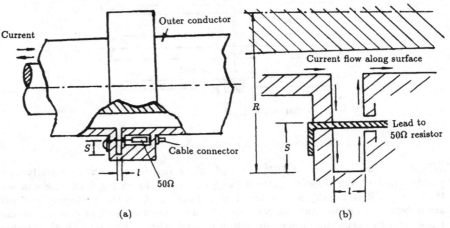

(a)   (b)

Fig.8-13 (a) Cross sectional view of the cavity current monitor; (b) An enlarged view of the cavity current monitor.

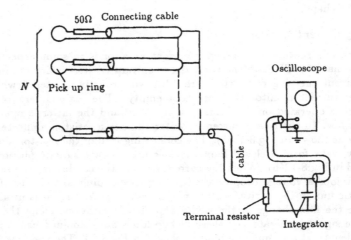

Fig.8-14 Equivalent circuit of the cavity current monitor.

where RC is the time constant of the integrator and N the number of loops employed. Fig.8-15 illustrates a typical example of using the cavity current monitor along with a capacitive voltage divider for the diagnostics of a high power pulse generator. The cavity current monitor is easy to construct and simple to operate and it can measure pulsed currents up to MA having risetimes in the subnanosecond range[14].

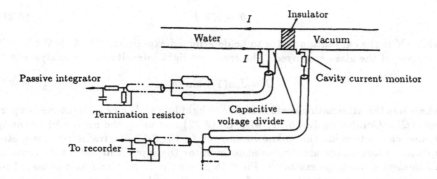

Fig.8-15 A typical example of using cavity current monitor and capacitive voltage divider for diagnostics of a high power pulse generator.

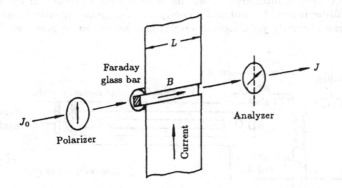

Fig.8-16 Schematic of the basic structure of the magneto-optical current sensor.

(d) Magneto-optical Current Sensor

Owing to their high electrical isolation and electromagnetic noise immunity, there is at present much interest in the use of magneto-optical sensors to measure pulsed currents generated in high power pulse generators. The basic structure of the magneto-optical current sensor built on the basis of the Faraday rotation is illustrated in Fig.8-16. A glass bar is inserted in a single-loop coil on the current-carrying conductor and is placed between a polarizer and analyzer having a $\pi/4$ angle between them. When a linearly polarized light of intensity J_0 is going through the glass, its intensity will change due to the rotation of its polarization vector. The rotated angle θ is related to the current I by the expression[15].

$$\theta = KV_cI \qquad (8\text{-}21)$$

where K is the calibration factor to be determined experimentally, V_c is the Verdit constant of the glass employed. The emergent light intensity at the analyzer is

$$J = J_0\alpha(1 + \sin 2\theta) \qquad (8\text{-}22)$$

where α is the attenuation coefficient due to light loss and can be determined experimentally. Combining Eq.(8-21) with Eq.(8-22) and using the measured intensity J, one can determine the unknown current I. Fig.8-17 illustrates two magneto-optical sensors which are employed to monitor the input and output currents of a Blumlein e-beam generator[15]. Fig.8-18 shows the respective result obtained by using the magneto-optical sensor A and the Rogowski coil A shown in Fig.8-17. From these results one can readily see the advantage of using the magneto-optical sensor for current measurements. There is practically no noise present whereas in the result obtained from the Rogowski coil there is severe noise present due to electromagnetic interference. As the environment surrounding the high power pulse facilities is usually electrically noisy, currents measured with magneto-optical sensors can generally yield cleaner results and hence can be more reliable.

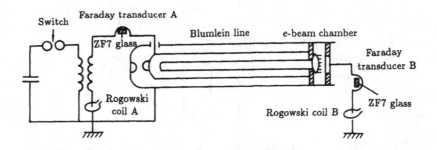

Fig.8-17 Fixed Faraday transducers for the input and output currents of the Blumlein electron-beam generator. A glass bar is inserted in a single-loop coil on the conductor of the Blumlein electron-beam generator to form the transducer [15].

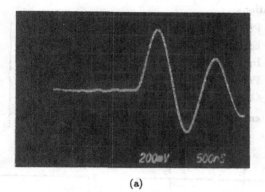

(a)

(b)

Fig.8-18 Current measured by (a) magneto-optical sensor A; (b) Rogowski coil A [15].

8-4 Measurements of Power, Energy and Other Quantities

(a) Quantities Inferable from Measurements of Current and Voltage

In high power pulse field, the determination and evaluation of many quantities other than the current and voltage are also essential and important. The most common quantities that are frequently needed for evaluation are the power, energy, impedance and perveance. In addition, the energy spectrum of charged particles generated from the system is also a piece of important information. Direct measurements of these quantities in general are not necessary and they can be inferred from the measured values of the current $I(t)$ and voltage $V(t)$ in conjunction with the following relations

Power	$P(t) = V(t)I(t)$
Energy	$\epsilon(t) = \int_0^t V(\tau)I(\tau)d\tau$
Impedance	$Z(t) = V(t)/I(t)$
Perveance	$F(t) = I(t)/[V(t)]^{3/2}$
Charged particle energy spectrum	$\Delta N(\epsilon) = \sum_i \dfrac{I_i(\epsilon)\Delta t_i}{e}$

$$(8\text{-}23)$$

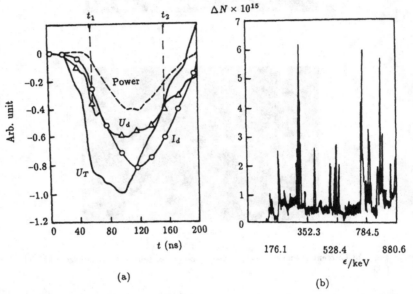

Fig.8-19 (a) Waveforms of the various outputs; (b) Electron energy spectrum obtained by Liu Junmin et al[16].

In the last expression, $\Delta N(\epsilon)$ represents the number of charged particles having energy ϵ and ϵ is defined by $\epsilon = eV(t)$. t_i is the time at which the charged particles having the same energy ϵ as illustrated in Fig.8-19(a) by t_1 and t_2. When these relations are applied to a device, such as an e-beam diode, all these quantities mentioned above must be referred to the diode only. In that case the diode voltage $V_d(t)$ needs to be inferred from other measurable quantities as $V_d(t)$ cannot be directly measured. The relation to be used for inferring $V_d(t)$ is

$$V_d(t) = V_T(t) - L\frac{dI_d(t)}{dt} \qquad (8\text{-}24)$$

where V_T is the output voltage of the transmission line connected to the diode, L is the diode inductance and I_d is the diode current. The diode voltage V_d so determined along with the diode power P deduced from Eq.(8-23) by Liu Junmin et al.[16] are shown in Fig.8-19(a). The electron energy spectrum deduced from the measured $I_d(t)$ and $V_d(t)$ along with the last relation given in Eq.(8-23) is shown in Fig.8-19(b). In the figure, ΔN represents the number of electrons having energy ϵ. $I_i(\epsilon)$ in Eq.(8-23) was obtained by eliminating t between $V_d(t)$ and $I_d(t)$ and setting $\epsilon = eV_d$.

Since the measured $V_d(t)$ and $I_d(t)$ are not analytic expressions of t, the procedure of eliminating t between $V_d(t)$ and $I_d(t)$ to obtain $I_i(\epsilon)$ has to be carried out by a numerical method. The digital recording technique has been found to be an effective means. In this method, appropriate monitors in conjunction with AD converters and microcomputer are employed to sample the data in digital form rather than continuous recording. The advantage of such a method is that the record of the measured values is available as a digital file and can be easily processed to reconstruct the profile of the measured quantities by means of some algorithms. The electron energy spectrum shown in Fig.8-19(b) was obtained by using this method. For more details about the application of digital recording techniques for the evaluation of high power pulse quantities, one may consult Ref.[17] or other texts.

Any measurement, no matter what method is employed, always contains some errors. In order to have an independent check of those quantities inferred from the measured values of current and voltage, some independent means to measure those quantities directly are necessary. Calorimeters and Faraday cups have been found to be useful devices for direct measurements of energies and currents of the charged particles generated in the diode. In the following sections, we shall respectively discuss these devices.

(b) Calorimeter

The basic principle of calorimeters may be described as follows. When the charged particles under study enter the calorimeter, the kinetic energy of the particles is absorbed by the absorbing body and the energy is converted into heat, thus the temperature of the absorbing body is raised. If the specific heat as a function of temperature is known, by measuring the temperature rise of the absorbing body, one can determine the total energy loss of the charged particles. The relation between the energy loss ϵ and the temperature rise of the absorbing body is

$$\epsilon = c\rho V \Delta T \qquad (8\text{-}25)$$

where c is the specific heat, ρ the density and V the volume of the absorbing body. The material employed for the absorbing body usually is graphite. There are a number of reasons that make the graphite attractive for such applications.

(I) Graphite has a relatively large stopping power for electrons[18]. For example, the thickness required to completely stop electrons of 1 MeV energy is only about

2.2 mm.

(II) Its back scattering coefficient for electrons is relatively small.

(III) The time required for reaching the state of thermal equilibrium is short. According to an estimate[19], it is about 10–15 sec.

The main sources that may introduce errors in the calorimetry measurements are: (1) back scattering of particles at the graphite surface, (2) energy losses due to radiation cooling and thermal conduction, (3) front surface spall and vaporization. Proper measures to reduce these effects are essential for achieving reliable measurements.

Basing on the requirements and applications, calorimeters of various forms have been designed by many investigators. Fig.8-20 is the schematic of a calorimeter designed to measure the total energy of the particle beams under study[20]. In this design the graphite body is made sufficiently thick (but not too thick to reduce the sensitivity) so that all the particle energy is absorbed. The temperature is monitored by a thermal couple which is electrically isolated from the calorimeter body. The graphite body in the calorimeter is electrically connected to the grounded plane to maintain its electrical neutrality. Diode energy obtained from such a device has been compared favorably with that calculated by integrating the product of the voltage and current[20]. In the design of such a calorimeter, proper care should be taken to minimize errors arising from the following effects. (1) some material on the graphite surface may be sputtered away due to intense bombardment of e-beams. One practical way to deal with it is to anneal the graphite plate at high enough temperatures or bombard the fresh surface of the graphite with an intense e-beam. (2) the output voltage of the thermocouple is in the millivolt range. To achieve reliable results, it is essential to use devices of high input impedance to make measurements, otherwise gross errors may be introduced.

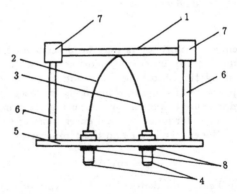

Fig.8-20 Basic structure of a graphite calorimeter [20]. 1–Graphite absorber; 2,3–Thermo-couple; 4–Signal output; 5–Support plate; 6–Support stick; 7–Thermo-insulator; 8–Mica wafer.

Fig.8-21 is the schematic of a calorimeter array designed to measure the spatial distribution of the particle energies[21]. A typical application of this type of devices is to check the uniformity of the electron beam source which is used to pump large size lasers. Such electron beam sources usually have a considerably large area of cross section whose uniformity is critical to the performance of the laser pumped. The calorimeter array employed in this application is usually placed behind the entrance window of the laser chamber so that the beam energy and uniformity can be directly checked.

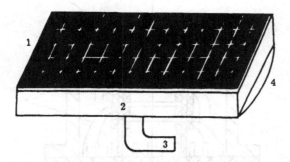

Fig.8-21 Basic structure of a calorimeter array [21]. 1, Thermo-plate; 2, Frame; 3, Electric output lead; 4, Back shielding cover.

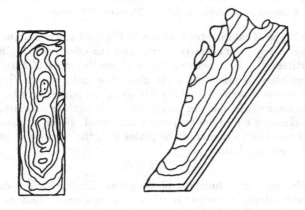

Fig.8-22 Simulated patterns of electron energy distribution [22].

One critical area which determines the performance of the calorimeter array is the contacts between the thermal couples and the graphite bodies in the array.

Improper contact will lead to false measurements. In order to improve the performance of the calorimeter array, some researcher[22] have used liquid Nitrogen temperature as the reference point of the thermal couples. The results so obtained for the spatial distribution of electron beam energy are shown in Fig.8-22. From these figures, one can see that the spatial distribution of the e-beam energies in this case is not quite uniform.

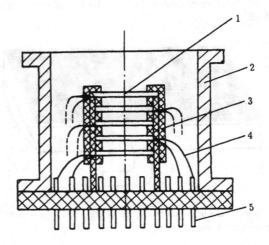

Fig.8-23 Basic structure of a multi-layer plate-calorimeter. 1, Plate absorber; 2, Cover; 3, Supports; 4, Thermo-couple; 5, Thermo-couple leads.

The third type of calorimeter is shown in Fig.8-23 and is designed to determine the maximum energy, the effective energy and the effective stopping range of the charged particles[22]. It is constructed with multi-pieces of thin plate to form the absorbing body. A selection of the absorbing material, the thickness and the number of plates is determined by the nature of the particles to be measured. For the measurements of the e-beam energy having moderate magnitudes Aluminum plates of thickness 0.1 mm are frequently employed. If the stopping range of the particle is R and the thickness of the plates is t, the number of plates N should satisfy the following relation

$$Nt \geq R \tag{8-26}$$

If the particles enter the absorbing plates perpendicularly and x denotes the particle penetration depth expressed in g/cm^2, by measuring the temperature rise in each plate, the quantity $H(x)$ which represents the heat absorbed per unit mass at the depth x can be determined. Further from the relation

$$I_B = \int_0^\infty H(x)dx \tag{8-27}$$

one can obtain the energy density of the particle beam which is denoted by I_B and expressed in Cal/cm^2. The depth x_m at which the value of $H(x)$ is zero represents the stopping range of the particles having the maximum energy. The ratio between I_B and $H(0)$ represents the effective stopping range, i.e.

$$x_e = \frac{I_B}{H(0)} \qquad (8\text{-}28)$$

From Eq.(8-28), one can determine the effective energy of the particle beam. With $I_B = 4.49$ J/cm^2, some previous authors[22] have determined the maximum and effective energies to be 0.583 MeV and 0.403 MeV, respectively.

When applying the calorimeter array, care must be taken to prevent the particle beam interacting with other mediums such as gas molecules. This is usually done by placing the entire set-up in an evacuated environment of approximately 10^{-4} Torr pressure. Under such conditions and with proper selection of the thermal couple materials, the device can reliably operate in a temperature range up to 1300°C.

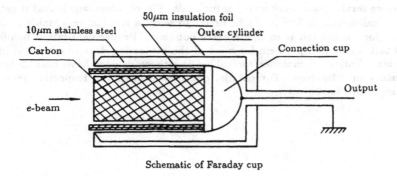

Schematic of Faraday cup

Fig.8-24 Schematic of Faraday cup [1].

(c) Faraday Cup

The basic structure of the Faraday Cup is similar to that of the first type of calorimeters described in the preceding section, except in this case there is no thermalcouple but with a shunt resistor connected between the graphite body and the ground. Fig.8-24 shows the schematic of a cylindrical Faraday cup in which a stainless steel cylinder serves as the shunt resistor[1]. When e-beam enters the graphite body, an EMF is generated thus producing a current flowing through the shunt resistor. This current intensity is proportional to the e-beam current injected into the graphite body. By measuring the voltage drop across the shunt resistance, one can determine the e-beam current. The maximum current I_m that a given Faraday cup can take is determined mainly by the breakdown strength of the foil of the insulation cylinder and its analytic form is

$$I_m = \frac{E_m D \ln(D/d)}{2R} \tag{8-29}$$

where E_m is the breakdown field of the insulation foil, D and d are the respective outer and inner radius, R is the shunt resistance. If the insulation foil has an $E_m \approx 0.2$ MV/cm, for a Faraday cup of 50 kA capacity, the required magnitudes of other parameters are approximately $D = 2.401$ cm, $d = 2.400$ cm and $R = 1.7$ mΩ. The time response of such a device is about 6 ns.

The Faraday cup is operated essentially on the same principle as that for the calorimeters except in the case of the calorimeter, emphasis is placed on the measurements of thermal effect whereas in the case of Faraday cups, measurements of the electrical effect is the goal, though both effects are present at the same time. Faraday cups can be used to measure either the e-beam or ion beam current[22]. They can also be operated with or without a bias. The main advantage of the Faraday cups is that they can measure the particle beam currents independently therefore serving as a check on other methods. The disadvantage is that it requires proper calibration before it can be used. To select a reliable, repeatable standard source for calibration is in many cases not easy. Proper calibration before use is in fact a common problem for most of the measuring devices discussed in this chapter. Various techniques for calibrating these devices have been developed by many investigators. For details, one may consult the respective references pertinent to the subject.

References of Chapter 8

[1] Liu Jinliang et al., *High power Laser and Particle Beams* (in Chinese), **5**, (1993), 629.

[2] S. Nakai et al., *Laser and Particle Beams*, **1** part 1 (1983), 29.

[3] M. Sato et al., *9th Int. Conf. on High Power Particle Beams*, II, (Washington D.C., 1992), 865.

[4] *Measurement of Electrical Quantities in Pulsed Power Systems*, Ed. by R. H. Mcknight and R. E. Hebner, *NBS special publication 628*, USA, (1982), 80.

[5] R. W. Stinnett et al., *9th Int. Conf. on High Power Particle Beams*, II, Washington D.C., (1992), 788.

[6] Yang Jin Ji et al., *Intense Impulse Current Technology* (in Chinese), (China Science Press, 1978).

[7] Wang Xiaojun and Yang Dawei, *High Power Laser and Particle Beams (in Chinese)*, **3**, No.3, (1991), 349.

[8] Neville W. Harris, in *Measurement of Electrical Quantities in Pulsed Power Systems*, Ed. by R. H. Mcknight and R. E. Hebner, NBS special publication 628, USA, (1982), 20.

[9] H. M. Herts and P. Thomson, *Rev. Sci. Instrum.*, **58**, (1987), 1660.

[10] G. Sower, *3rd IEEE Pulsed Power Conf.*, (1981), 189.

[11] Yu Cunyi et al., *The Scientific and Technical Report* (in Chinese), **No.93-045**, Xi'an Jiaotong University, (1993).

[12] M. S. Di Capua, *High Speed Magnetic Field and Current Measurements*, in NATO ASI Series E, Appl. Sci No. 108, (Plenum, 1986), 225.

[13] J. Shannon, E. Chu, R. Richardson, M. Wilkinson and C. Trivelpiece, in *Measuremet of Electrical Quantities in Pulsed Power Systems*, Ed. by R. H. Mcknight and R. E. Hebner, NBS special publication 628, (1982), 289.

[14] *Operation and Maintenance Instructrons for POCO 1-100 pulseline and support equipment*, Maxwell, MLR-1759, (1984), 6-1.

[15] S. T. Pai et al., *Sensors and Actuators A*, **35**, (1992), 107.

[16] Liu Junmin et al., *High Power Laser and Particle Beams* (in Chinese), **4**, No.3, (1992), 343.

[17] R. Malewski et al., in *Measurement of Electrical Quantities in Pulsed Power Systems*, Ed. by R. H. Mcknight and R. E. Hebner, NBS special publication 628, (1982), 341.

[18] L. Page, E. Bertel, *Atom Data*, **4**(1), (1972), 1.

[19] N. I. Gaponenko, *Pribory i Tekhnika Eksperimenta*, **3**, (1976), 47.

[20] Fan Yajun, *4th Chinese National Cinference on High Power Pulse (in Chinese)*, Xi'an, 1990. E. Nalting et al., in *Measurement of Electrical Quantities in Pulsed Power Systems*, Ed. by R. H. Mcknight and R. E. Hebner, NBS Special Publication 628, (1982), 118.

[21] Feng Qi et al., *High Power Laser and Particle Beams (in Chinese)*, **3**, No.3, (1991), 405.

[22] Han quangyuan et al., *3rd Chinese National Conference on High Power Particle Beams (in Chinese)*, Chende, 1988.

APPENDIX
List of Symbols

A — area, cross section.

B — magnetic induction (flux density).

C — capacitance.

C_l — load capacitance.

c — speed of light, specific heat.

d — gap spacing.

E — electric field.

e — electrons, electronic charge.

F — force.

f — frequency, field enhancement factor.

G — conductance, gain.

g — gain.

H — magnetic field.

h — height.

I — diode current.

I_e — electron current.

I_i — ion current.

i — total current.

J — electric current density, light intensity.

j — $\sqrt{-1}$

K — thermal conductivity.

k — field constant.

L — inductance.

L_l — load inductance.

L_t — time-dependent inductance.

l — length.

M — ion mass, mutual inductance.

m_0 — electron rest mass.

N — channel number.

n — electron density.

P — power, pressure, energy density.

R — resistance, radius.

R_l — load resistance.

R_t — time-dependent resistance.

r — radius.

S — switch.

T — pulse width, temperature.

t — time.
t_r — rise time.
V — voltage.
V_b — breakdown voltage.
v — velocity.
Z — impedance.
Z_l — load impedance.
α — absorption coefficient, Townsend 1st ionization coefficient.
β — ratio of v/c.
γ — relativistic factor.
δ — electron sheath.
η — efficiency, electric resistivity.
ϵ — energy, electric permittivity.
ϵ_f — Fermi energy.
λ — wavelength.
μ — magnetic permeability.
ρ — density, reflection coefficient.
τ — time constant.
ϕ — work function.
θ — deflection angle.
ψ — magnetic stream function.
ω — angular frequency.

Units Frequently Used and Their Magnitude

$1 \text{ MA} = 10^6$ Ampere
$1 \text{ MV} = 10^6$ Volt
$1 \text{ MJ} = 10^6$ Joule
$1 \text{ MPa} = 10$ Atmosphere
$1 \text{ TW} = 10^{12}$ Watt
$1 \text{ nm} = 10^{-9}$ meter
$1 \text{ ns} = 10^{-9}$ sec
$1 \text{ L} = 10^3 \text{ cm}^3$
$1 \text{ R} = 2.58 \times 10^{-4}$ C/kg

Index